AF412923

Power Transformer Handbook

Power Transformer Handbook

Edited by Bernard Hochart

Alsthom Transformer Division,
Saint-Ouen, France

First English edition
translated from the French by
C. E. Davison BSc(Eng), FICE, FIWES, MASCE

Butterworth-Heinemann
Linacre House, Jordan Hill, Oxford OX2 8DP
A division of Reed Educational and Professional Publishing Ltd.

A member of Reed Elsevier plc group

OXFORD BOSTON JOHANNESBURG
MELBOURNE NEW DELHI SINGAPORE

First published 1982
English edition 1987
  Reprinted 1989
  Reprinted 1999

ISBN 0-408-02590-5

Printed in India by Replika Press Pvt Ltd, 100% EOU, Delhi-110 040

The book is exclusively distributed in India by Aditya Book Pvt Ltd,
2/37 Sant Vihar Gali, Ansari Road, Daryaganj, New Delhi 110 002

Preface

This book on power transformers encapsulates the improvements in design and construction resulting from the efforts of successive generations of engineers and technicians, representing more than 60 years of design calculation, fundamental research and testing.

It presents this information in an accessible form for those responsible for selecting, buying, installing, operating and repairing transformers and to help those responsible for instructing future generations of engineers.

Throughout its 20 chapters, the buyer will find information which is necessary for making the correct choice of transformer for his own requirements. The erector will learn the rules for good operation of the equipment, and the user will see how to use transformers without prejudicing the good operation of a distribution system.

In the field of power transformers progress is rarely spectacular. It results from continuing small improvements in reliability and performance to control costs.

The two final chapters, which are more specialized, deal with certain applications in detail, particularly the recovery of energy, and other topical concerns.

The study of the supply of power to an industrial installation or to a power transmission and distribution system is a major undertaking which is carried out over a long period of time. This book with its pragmatic approach to the problems that arise can help the reader to reduce the risks of error or assist him in communicating with the manufacturer.

Contributors

Jacques Bassot
Sylvain Cagnioux
Robert Dides
Maurice Gallay
Jean-Marc Gorria
Jacques Lemaire
André Malandain
Ernest Meier
Daniel Périé
Claude Piccon
Jean Poittevin
Paul Roth

Transformers illustrated in this book were manufactured by Alsthom in its Le Havre and Saint-Ouen factories, and are installed world-wide.

Contents

1 Recent advances

A casual observer of transformer technology might think that little advance has been made for a number of years.

While it is true that the principles of electromagnetism on which the technology is based have not changed, the improvements made in magnetic materials, conductors and insulating materials, and the way in which they are used, have enabled service voltages of 800 kV to be attained with increased reliability and units of over 1600 MVA to be constructed in spite of the ever increasing constraints on transporting large indivisible pieces of equipment.

Performances at ultra high voltage and very high power are the best evidence of a high technical and technological mastery, but it should not be forgotten that transformers of all powers and voltages are benefitting from the progress that has been made.

ADVANCES IN SPECIFICATIONS

The advances in specifications over the years have in general tended to reduce the overall cost of transformers (both capital cost and running cost). Three examples will illustrate this.

Loss level

With a continuing demand for economy it is inevitable that improvements should be expected in transformer performance. Customers insist on a

Figure 1.1. Evacuation energy transformer, 760 MVA, 20/405 kV three phase

reduction in losses and compare tenders by adding together the capital cost of the transformer and the capitalized cost of their losses. This policy has led to slight oversizing of the core and windings to reduce the losses in service.

Use of a Guide to Loading

The use of the IEC Guide to Loading 354 enables the cost of the core and windings to be reduced when the load is discontinuous.

Reduction in level of insulation

Increasing use of modern lightning arrestors has led to a reduction in the level of insulation required against lightning. For example, 220 kV systems have had their insulation level reduced successively from 1050 kV to 900 kV, and in certain cases the lowest level is 650 kV.

This has been accompanied by a reduction in the cost of the equipment and an increase in the operating security.

RECENT ADVANCES MADE BY SUPPLIERS OF MATERIALS

As the cost of materials forms about 50 per cent of the sale price, any progress made by suppliers in reducing costs makes an appreciable difference to the equipment costs. Three examples demonstrate this.

Reduced loss magnetic laminations

Improved core steel, using new processes, is now available on the market enabling substantial technical gains to be made.

Losses

Since its appearance on the market, grain-oriented steel has made it possible to achieve appreciable increases in the working flux density of transformer cores. See Figures 1.2 and 1.3.

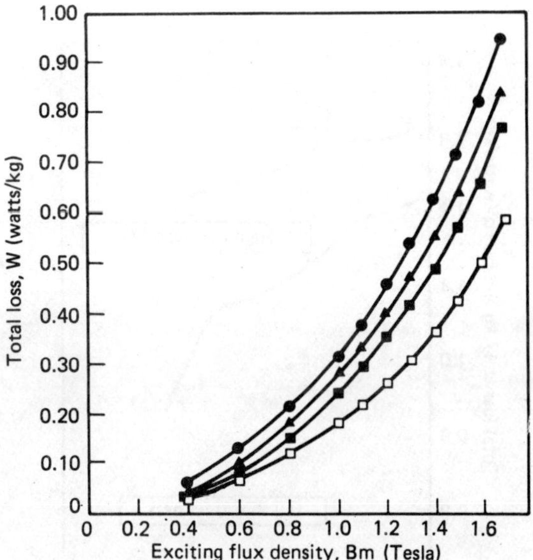

Figure 1.2. Correlation between losses and thickness

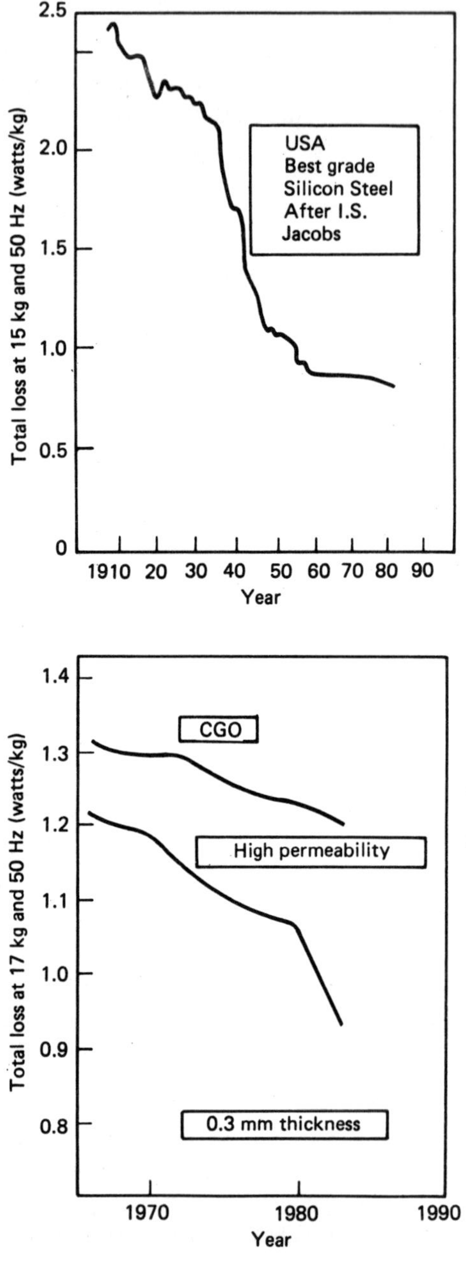

Figure 1.3. Improvement of best-grade commercial grain-oriented silicon steel

An analysis of the losses showed that 50 per cent were due to eddy currents, and were therefore related to the thickness of the laminations, and 50 per cent were due to hysteresis.

Steel manufacturers have progressively reduced the lamination thickness from 0.35 mm to 0.23 mm and future expectations are for laminations only 0.15 mm thick.

The manufacturing processes and the metallurgy of these sheets have both been improved (particularly in grain orientation, though this leads to higher losses when the magnetic flux is at right-angles to the direction of rolling) leading to improved performance (low losses).

Steel manufacturers now offer sheet with a range of characteristics from which the manufacturer can choose to suit his method of construction and the required performance.

Noise

The noise generated by the magnetic sheets is due to the phenomenon of magnetostriction in which the amplitude is related to the induction. With certain types of sheets (Hib) it is possible to reduce the inherent noise level of the sheet by several decibels without changing the induction, independent of the method of constructing the magnetic circuit which is itself an important factor in determining the level of noise from a transformer.

Transposed conductors

An important source of loss in the windings is unequal current distribution in the conductors (skin effect).

To reduce the additional losses due to this phenomenon it is current practice to sub-divide conductors. Manufacturers of insulated wire have improved the processes of manufacturing multiple-wire transposed conductors (Figure 1.4). The number of wires in a transposed conductor can, today, exceed 80. The hardness of the wire material has been increased from 12 to 24 kg/mm^2 for an elongation of 0.1 per cent. These cores can be coated with epoxy resin which, after polymerization, increases the mechanical strength of the cable.

Connection with SF6 insulated terminals

The use of these conductors is general for very high currents and/or when reduced losses are specified. They reduce the coil winding time and consequently increase the factory production capacity.

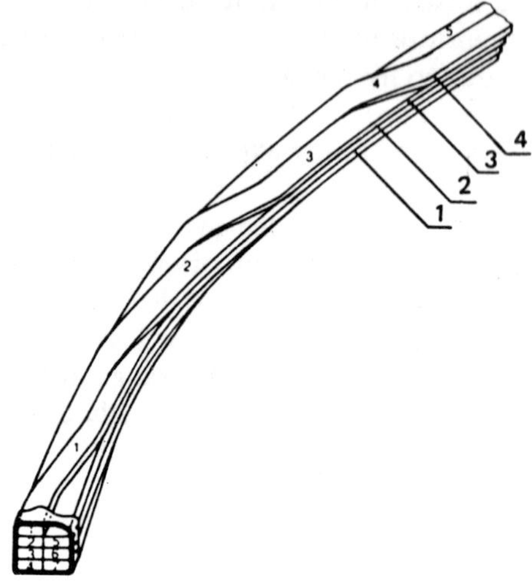

Figure 1.4. Conductor with several insulated and transposed wires

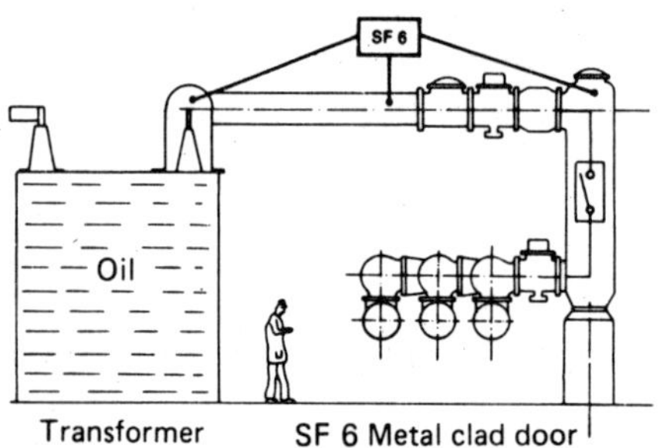

Figure 1.5. Connection between transformer and metal clad substation

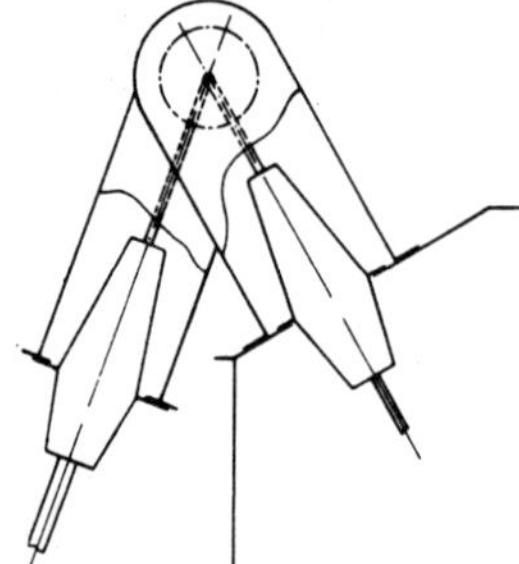

Figure 1.6. Connection between transformer and cable via SF6 filled chamber

The connection of a transformer to a high-voltage metal-clad substation can be made without using cables (see Figure 1.5).

A fourth example of cost saving is where a fault with a cable joint box must be avoided, and some customers require the connection to be made by means of an oil chamber–SF6 compartment–cable (see Figure 1.6).

ADVANCES IN CALCULATION

The development of computers has enabled field plots obtained by means of electrolytic tanks to be replaced by field plots calculated for both

electrostatic fields and electromagnetic fields. From this a better knowledge has been obtained of dielectric stresses due to over-voltages and
electrodynamic stresses due to short-circuit currents.

Electromagnetic models developed in the 1960s can now be replaced by
computer simulations which have provided a more precise knowledge of
the distribution of voltages along windings when they are subjected to
lightning strokes and switching over voltages.

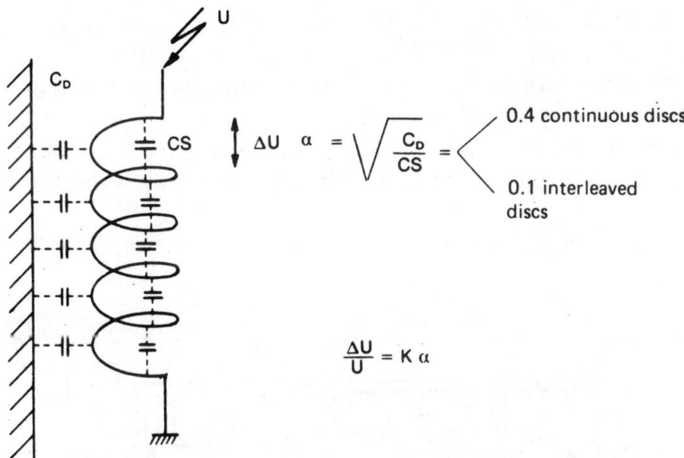

$$\Delta U \quad \alpha = \sqrt{\frac{C_D}{CS}} = \begin{cases} 0.4 \text{ continuous discs} \\ 0.1 \text{ interleaved discs} \end{cases}$$

$$\frac{\Delta U}{U} = K\alpha$$

Figure 1.7. Interleaving reduces axial stresses due to voltage impulses

Uniformity of dielectric stresses

Concerned with finding an insulated winding construction as uniform as
possible, whatever the electrical stresses it would be subjected to,
designers have thought up arrangements of the turns relative to one
another aimed at reducing the ratio of shunt capacitance to series capacitance. By shunt capacitance is meant the distributed capacitance related to
its environment, that is to magnetic core circuit, tank or the adjacent
winding. See Figure 1.7.

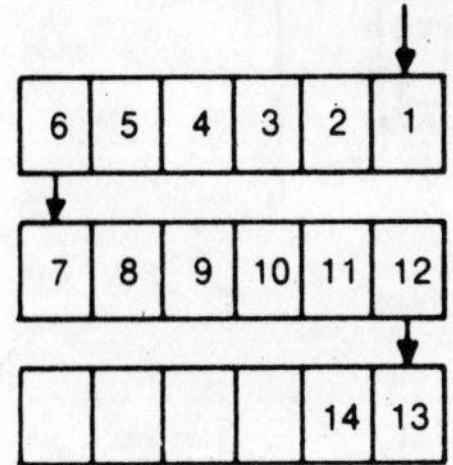

Figure 1.8. Continuous-disc winding

By series capacitance is meant the capacitance either between turns or between groups of turns.

This search for uniformity of stresses is all the more important as the voltages in the windings increase, but it is not independent of the increase in unit power as the voltage distribution tends to improve with increased rating.

In manufacture, therefore, we have *continuous disc windings* (see Figure 1.8) and *continuous disc windings with floating potential shields* (see Figure 1.9). These shields generally consist of non-current-carrying turns wound between the actual current-carrying turns, and are interconnected between adjacent coils. The voltage between the shield and the adjacent turns is higher than the turn-to-turn voltage, so that the energy stored in the series capacitance is very much increased.

We also have *interleaved disc windings* (see Figure 1.10) and *layer-type windings* (Figure 1.11). The arrangement of interleaved discs is similar to

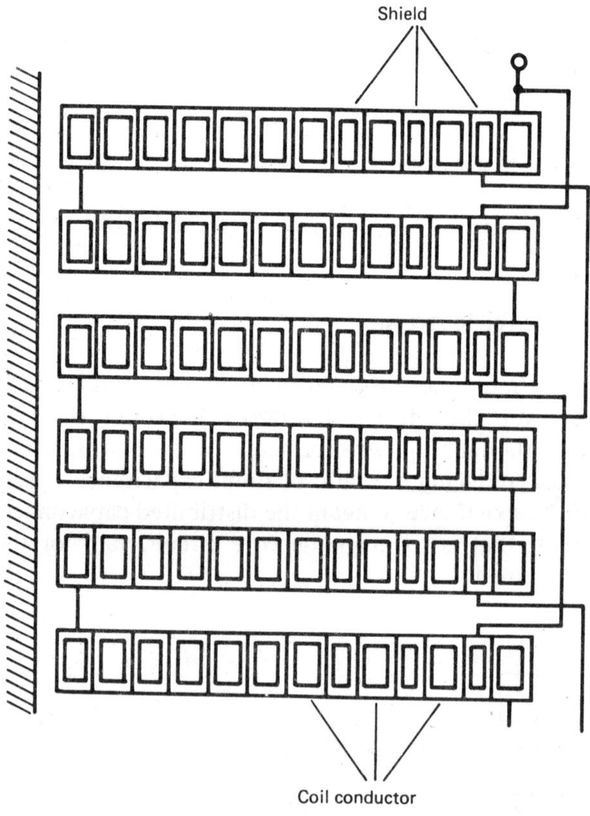

Figure 1.9. Shield winding

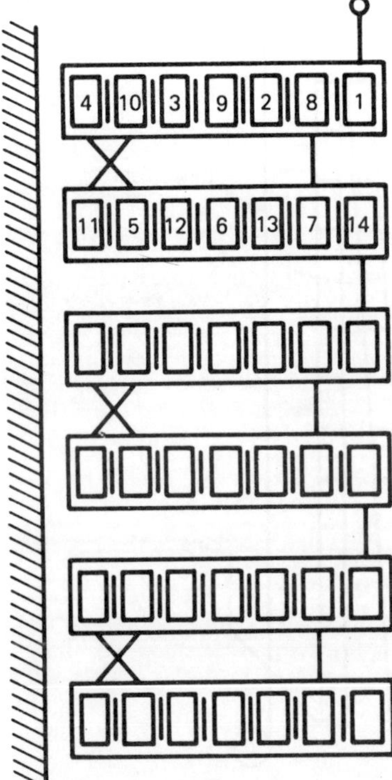

Figure 1.10. Interleaved winding.

the continous discs with shields, but the increase in capacitive energy is obtained by systematically interleaving the turns of one coil with those of the following. For reasons of economy, in terms of the 'space-factor', these turns, which act as the capacitive screen in the preceding case, now have current flowing through them and therefore become the turns of the winding itself.

In layer-type windings, the adjacent turns are no longer arranged radially but axially and the winding consists of a certain number of concentric layers separated by oil paper insulation.

While in windings consisting of disc coils an increase in series capacitance is sought, the arrangement of the windings in concentric layers tends to decrease the shunt capacitance, the layers forming coaxial capacitances being connected in series to earth. For very high voltages, the distribution of voltage is improved even more by the use of electrostatic screens arranged concentrically to the layers and connected generally either to the line or to neutral, or sometimes to both these extremities.

This type of winding is particularly useful in the construction of auto-transformers interconnecting two very high voltage systems.

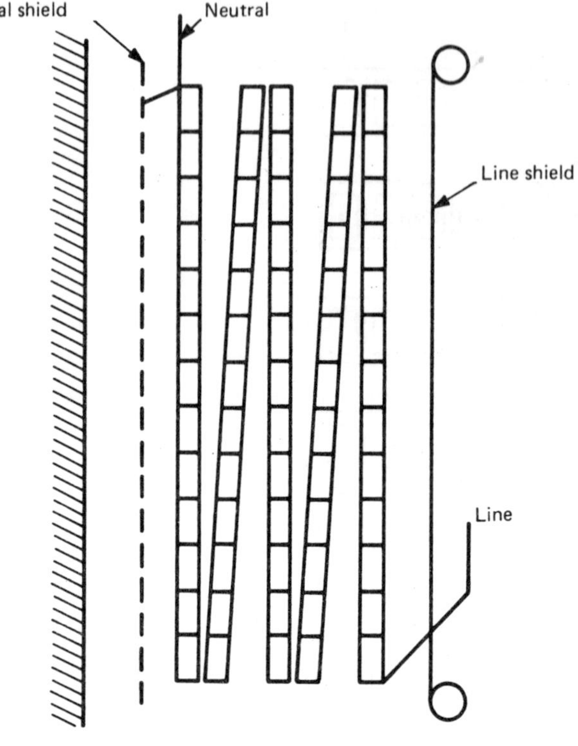

Figure 1.11. Layer winding

ADVANCES IN DESIGN

A transformer consists of an assembly of independent elements each of which has been the subject of development. Some examples follow.

Magnetic circuit

The development of grain-oriented steel has led to elaborate methods of construction in which the magnetic flux follows as near as possible the preferential direction in the sheets (see Figure 1.12). This has resulted in the form of construction known as a 'mitred joint', in between the legs and yokes of the core.

Core bolts have been eliminated in assembly work, and are replaced by glass fibre bands impregnated with thermal setting resin.

Figure 1.12. Core of a 70 MVA, 227 kV, three-phase unit

Windings

Low voltage windings

The technique depends principally on the current. For low power, the windings may consist of aluminium or copper foil wound with a turn to turn insulating foil. For medium power, they consist of one or more layers of transposed cable, separated by oil ducts concentric with the layers (Figure 1.13). For high power, hence high currents, each turn in these layers will consist of several transposed conductors in parallel and these will possibly be separated by horizontal cooling ducts. This type of winding is also called a helical winding (Figure 1.14).

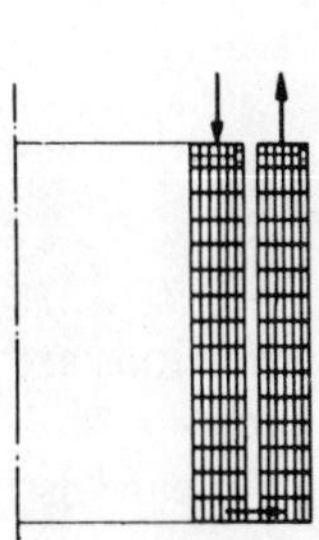

Figure 1.13. Winding consisting of two superimposed layers

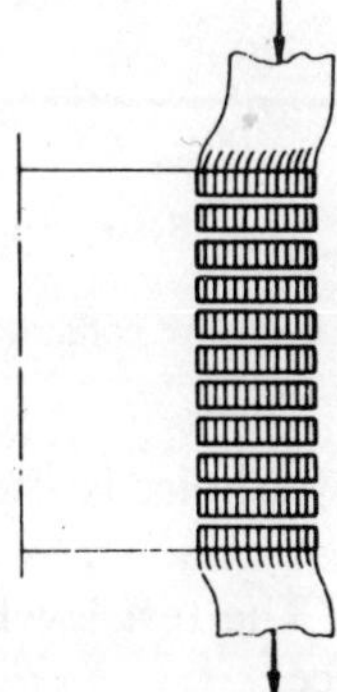

Figure 1.14. Continuous helix in flat parallel wires

High voltage windings

The choice of the type of winding depends on several factors:

Voltage – Voltages up to about 130 kV may not require the inclusion of a shield, but higher voltages generally do.

Power rating – The higher the power, the higher is the point at which a shield is needed. Figure 1.15 gives an idea of the choices possible as a function of these two parameters.

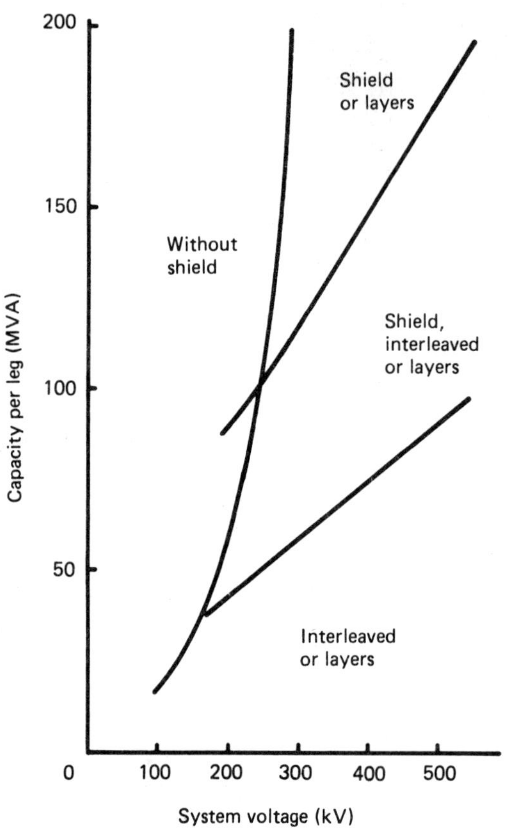

Figure 1.15. Choices of construction of high voltage windings

Winding connections – Choice is affected by whether the connections are star or delta.

Level of insulation – The test levels between line and neutral can also lead to a different choice.

Ratio of interconnected voltages – In the case of autotransformers, this will lead to the choice of one winding type rather than another.

Axial support of windings

The axial support of the windings has been made almost independent of the magnetic circuit. It is effected by a beam assembly connected together by tie-bars and clamping plates put under pressure by clamping jacks. See Figure 1.16.

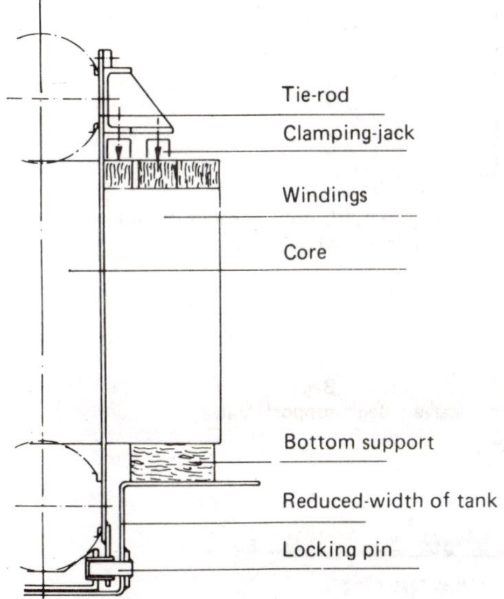

Figure 1.16. Axial clamping of windings

Radial support of internal windings

The internal winding (usually low voltage) is wound on a support cylinder. Adjustable wedges between this cylinder and the magnetic circuit increases the radial rigidity of this winding.

Tank fabrication

Advances in tank fabrication have led principally to a reduction in volume, giving an appreciable saving in quantity of insulating oil. The lower yoke rests in a narrow trough and the side wall on the HV terminal side is usually inclined (Figure 1.17).

Membrane reservoir

Techniques are now available for the economic and reliable separation of air from oil in the conservator (Figure 1.18).

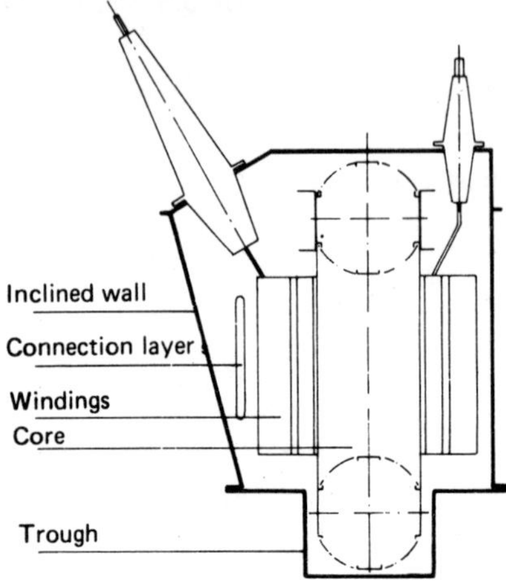

Figure 1.17. Arrangement of tank to reduce insulation oil quantity

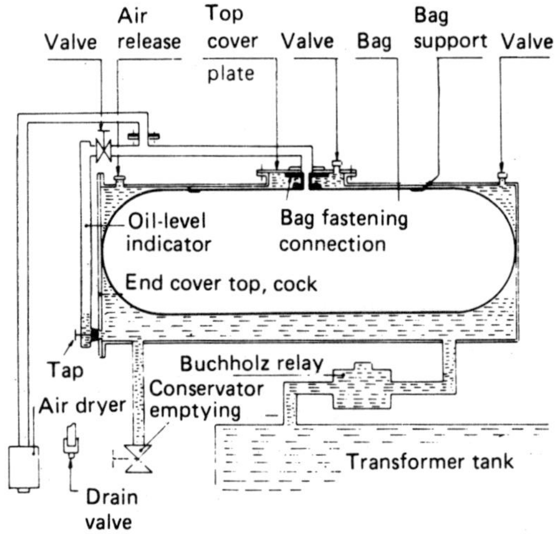

Figure 1.18. Arrangement of conservator with rubber bag

ADVANCES IN MANUFACTURE

Large transformers can only be constructed economically with high capacity methods of production including appropriate methods of lifting, equipment and machine tools, and testing equipment.

Among recent developments is a machine for cropping magnetic laminations.

Machine for cropping magnetic laminations

This is shown in Figure 1.19, and has the following characteristics:

maximum width of strip to be cut: 1000 mm;
length of sheeting cut: 5000 mm;
controlled by a machine process controller, which is itself programmed by a punched tape;
cropped laminations sorted and pre-stacked at two levels, each level consisting of four compartments.

This machine achieves a high degree of dimensional precision. A big reduction in the air gaps has been made possible in the core to leg joints with a consequent improvement in the no-load current and loss, and the noise.

Figure 1.19. Cropping machine for core plates

Magnetic circuit stacking unit

Magnetic circuits up to 300 tonnes can be stacked on this unit. An ergonomic study has shown the advantages of stacking circuits up to 50 tonnes less the upper yoke. The size of the equipment makes it uneconomic to use this method for very large transformers.

Winding machine

The largest machines enable windings of the following characteristics to be constructed:

weight: 30 tonnes;
length: 3000 mm on a vertical machine, 3450 mm on a horizontal machine;
diameter: 3000 mm on a vertical machine, 2600 mm on a horizontal machine.

Press

An 800 tonne press enables the windings of the largest transformer to be stabilized using vacuum and heating cycles. Hydraulic jacks maintain a constant axial pressure throughout these cycles. Compared with the older system of clamping by plates and springs, the advantage is gained in the total duration of these operations.

Figure 1.20. Press assembly stabilizing all six windings of a three-phase transformer

Vapour phase drying

This has the advantages of very effective cleaning of the core and windings, and more rigorous and rapid drying, with less ageing of the insulating material.

Figure 1.21. Autoclave for vapour-phase drying treatment

Shot blasting and painting shop

Modern facilities enable the biggest transformers (760 MVA) to be treated.

ADVANCES IN QUALITY CONTROL AND TESTING

The most recent advances have been made in the following areas:

Testing for, and location of, partial discharges on completion of manufacture

This makes it possible, in 80 per cent of cases, to locate with sufficient precision any manufacturing defect that is likely to cause partial discharges in service.

Monitoring hot spots

Temperature rise tests which were aimed at checking the average temperature rise of the windings have been complemented by measurements of the

Figure 1.22. Transformer tank painting booth

changes of the gases dissolved in the oil. These measurements enable the presence of hot spots to be detected, but they do not enable such spots to be precisely located. Infra-red photography also detects possible hot spots on the transformer tank.

Verification of the short-circuit strength

Numerous transformers from 1 to 600 MVA and from 20 to 420 kV have been subjected to short-circuit tests in recent years (1 to 24 short circuits per phase). These very costly tests make it possible to improve the reliability of the basis of construction for all the equipment.

ADVANCES IN SITE ERECTION

Progress in site erection has been mainly in the study of erection methods, and in the supply of modern lifting equipment and equipment for treating and storing the oil.

ADVANCES IN PRODUCTION MANAGEMENT

To remain competitive in the modern world it is no longer sufficient to be ahead technically. It is necessary to survive on ever reducing margins. This is possible only by increasing the production capacity of the works. From design office, through the workshop to the test shop, the manufacturing time must be reduced. Numerous steps have been undertaken to achieve this, among which the most effective have been the planning and forecasting of the resources (staff, tools, working areas) extending throughout all the departments which contribute to the manufacture of transformers. This has led to a reduction in manufacturing time of 15 per cent.

2 Voltage regulation

VOLTAGE VARIATION IN DISTRIBUTION SYSTEMS

The power demands made on a distribution system by consumers are constantly changing, causing variations of voltage with time at various points in the system due to voltage drop. There are two types of variation:

short-term (lasting from a fraction of a second up to ten seconds or so) due to equipment being started up, fluctuations of power (electric arc furnaces) and operation of the system. They cause fluctuations in lighting, and disturbances in the operation of electronic equipment, motors, etc.
long-term (duration exceeding ten seconds and up to several hours) due to progressive start up or shut down of numerous pieces of equipment of low power and to changes in operating requirements. The nuisance to consumers derives from the average voltage being too much above or below the nominal operating voltage.

Only slow variations of voltage which can be controlled by transformer tap changing are considered here.

VOLTAGE REQUIREMENTS OF INDUSTRIAL EQUIPMENT

Equipment using electrical energy has optimum operating conditions at a predetermined voltage called the 'rated voltage' (U_n). As, in most cases, it is not possible to supply continuously at this voltage, the equipment is

Figure 2.1. Core and windings of a 100 MVA transformer, for interconnection between 220 kV and 90 kV systems. On-load tap changing of ±15 per cent on the 220 kV winding

designed in accordance with the appropriate standards to have a satisfactory range of operation around this value.

For example:

Electric motors: rated power from 95 to 105 per cent of U_n.
Transformers: voltage variation from 95 to 105 per cent of U_n.
Capacitors: prolonged operation at 110 per cent of U_n.
Portable tools: 10 starts at 85 and 110 per cent of U_n.

Electrical equipment for machine tools: starting and operation between 95 and 105 per cent of U_n.
Electromagnets: operation between 90 and 105 per cent of U_n.
Fluorescent lamps: operation between 90 and 110 per cent of U_n.

In fact, the operation of some equipment becomes poor or impossible beyond certain limits. The luminous flux of a lamp varies with the cube of the voltage, and a 5 per cent increase in voltage reduces its life by a half. Below a certain level, fluorescent lamps only light up with difficulty.

A motor supplied with a low voltage will not produce its rated torque and its windings will overheat. With too high a voltage, the magnetic circuit becomes saturated and overheats.

Some equipment, such as computers, requires a virtually constant voltage without interruption. The only safeguard in these cases is an independent power supply specially provided for this purpose.

MATHEMATICAL EXPRESSION FOR VOLTAGE IRREGULARITY

The drawbacks of a supply being at a voltage different from that of the nominal voltage have been the subject of theoretical studies, with a view to determining the best conditions for voltage regulation in distribution systems. These studies, based on statistical methods, are to a large extent applicable to industrial systems supplied from a transformer substation.

The general characteristics of electrical equipment are such that the drop in performance by supplying a piece of equipment at a voltage different from the rated voltage increases with the square of the difference between

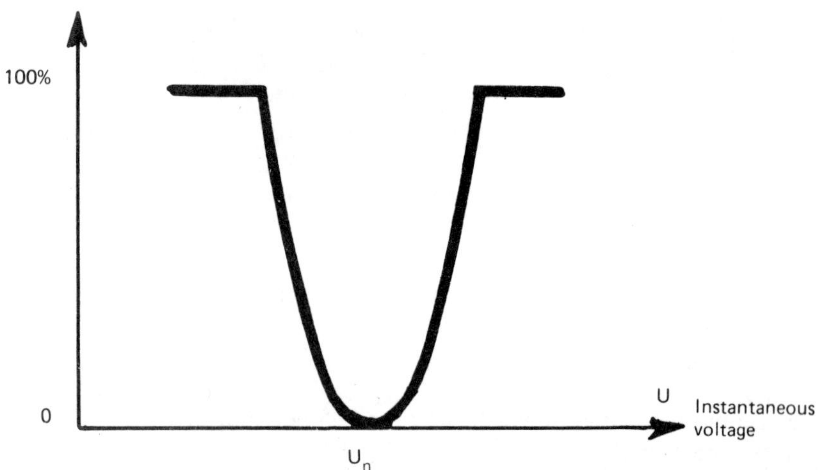

Figure 2.2. A single item of equipment

the two. For large voltage differences operation of the equipment is impossible (Figure 2.2).

With the voltage varying all the time, the irregularity over a period $t_1 - t_2$ is expressed by the average value of the square of the differences in percentage between the instantaneous voltage U and the rated voltage U_n.

$$I_t = \frac{1}{t_2 - t_1} \int_{t_1}^{t_2} V^2(t)\mathrm{d}t = \overline{V^2} = \overline{V}^2 + \sigma_v^2$$

where

$$
\begin{aligned}
V &= 100(U - U_n)/U_n, \\
\overline{V} &= \text{average value of } V \text{ between } t_1 \text{ and } t_2, \\
\overline{V^2} &= \text{average value of } V^2 \text{ between } t_1 \text{ and } t_2, \\
\sigma_v &= \sqrt{(\overline{V^2} - \overline{V}^2)} = \text{standard deviation } V.
\end{aligned}
$$

There are two relevant factors – the average variation from the rated voltage, and the fluctuation above and below this.

It is easy to compensate for the first (by adjustment of the transformation ratio of the supply transformer). The second is much more expensive to deal with because of the cost of fitting on-load voltage regulation equipment, and so is remedied only in real necessity. For several pieces of equipment with different characteristics supplied from the same source, the superimposed individual diagrams form an irregular curve (Figure 2.3a) in which the minimum is not zero. It only becomes zero if all the equipment has the same nominal voltage (Figure 2.3b).

If a group of electrical equipment in a factory is considered, such as motors, electric furnaces, machines, etc. a more elaborate concept is arrived at in which the irregularity of the overall voltage is the sum of the irregularities in the individual voltages weighted by the individual power demands at each instant. Special voltmeters have been produced which directly give this value in (per cent)^2 × kWh, thus expressing the voltage irregularity and, to a certain extent, its drawbacks.

CORRECTION OF VOLTAGE VARIATIONS IN A FACTORY DISTRIBUTION SYSTEM

An industrial system is usually supplied from a substation connected to the grid system at medium voltage (15, 20, 30 kV) or high voltage (63, 90, 225 kV). The substation consists of an MV/MV transformer if it is connected to medium voltage or an HV/MV transformer if connected to high voltage (Figure 2.4).

The power is distributed at medium voltage, 3–20 kV, according to the various requirements, to distribution points around the factory situated

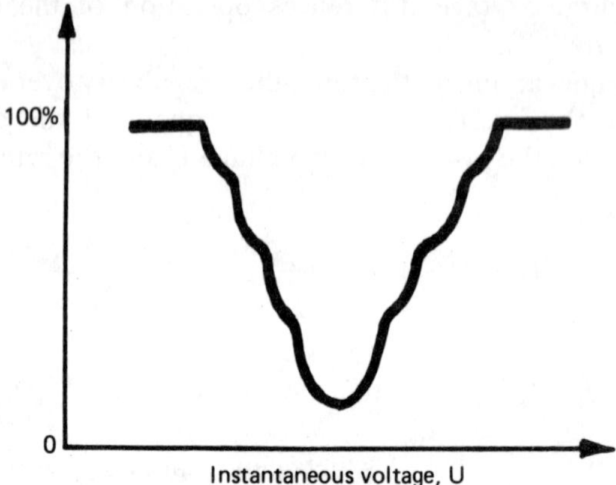

Figure 2.3a. Several items of equipment with different rated voltages

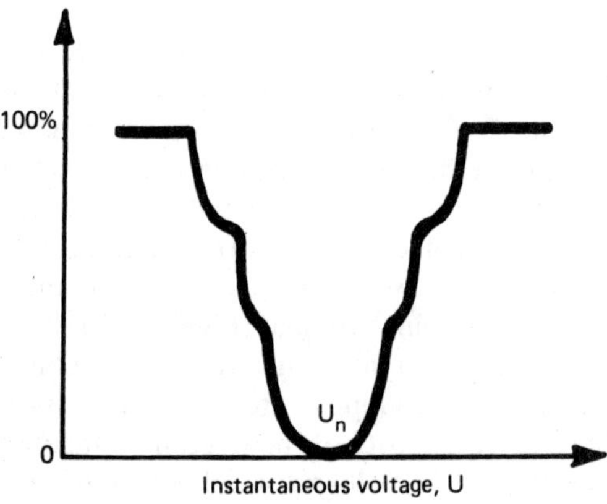

Figure 2.3b. Several items of equipment with the same rated voltages

close to the principal centres of consumption, where the voltage is reduced to 220 and 380 V. Equipment such as large motors are supplied directly at medium voltage.

Auxiliary circuits such as lighting generally have their own circuits, often from an MV/LV transformer.

The voltage regulation can be made by means of off-circuit tappings on MV/LV transformers (or HV/MV), by capacitors on the medium or low voltage systems, or by on-load regulation of HV/MV transformers. In all cases the regulation is based on the average value of the voltage, and in the last case the fluctuations are also remedied.

On-load regulation is essential for equipment that cannot accept voltage

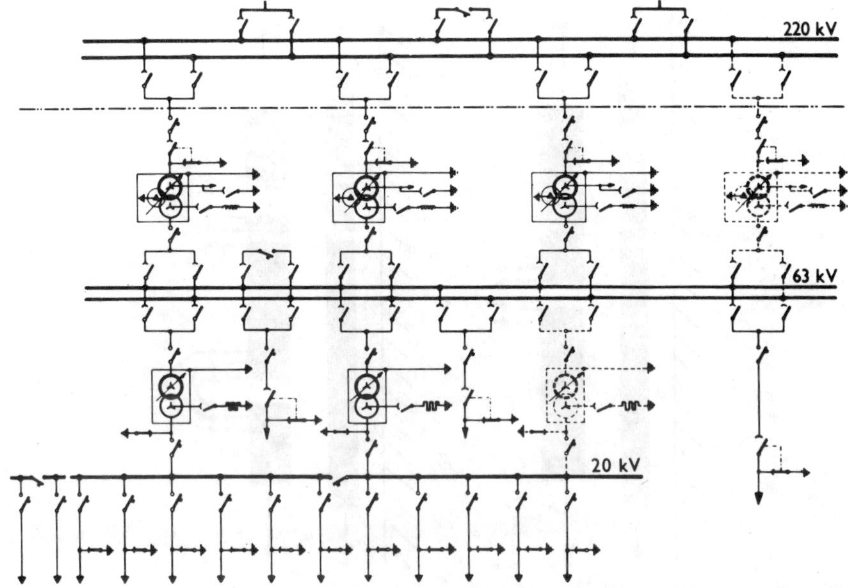

Figure 2.4. An example of an HV/MV substation

fluctuations. Being an important price element of a transformer, it is generally only used for power ratings of 10 MVA or higher with the supply usually at high voltage.

Capacitors are used to compensate the reactive current, resulting in a reduction of voltage drops.

The normal process of voltage regulation starts at the supply substation. If the supply is at medium voltage (already regulated at some previous point) through an MV/MV transformer, it is possible to adjust the average voltage of the factory system by operating the off-circuit taps.

If the supply is at high voltage, the on-load regulation of the HV/MV transformer enables the average voltage of the internal system to be adjusted at the same time as reducing the effects of voltage fluctuations upstream and downstream. Finally by means of the off-circuit taps on the MV/LV transformers in the workshop distribution points, the low voltage can be centred on its rated value at the different points of use.

TRANSFORMERS WITH OFF-CIRCUIT TAPS

Adjustment of the transformation ratio requires the introduction of tappings on one of the windings, corresponding to the required steps in regulation. In the case of off-circuit regulation, the change from one

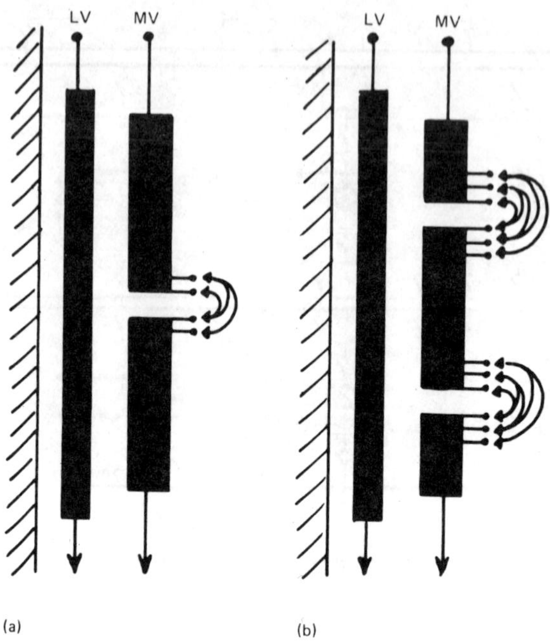

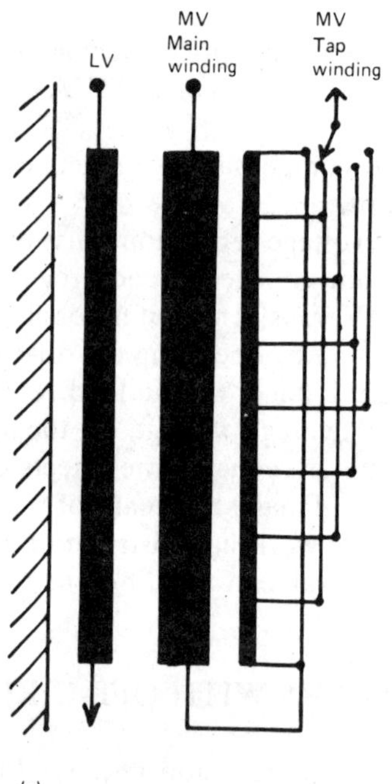

Figure 2.5. (a) Small tapping range (b) Medium tapping range (c) Wide tapping range

tapping to another is carried out by means of a switch which is designed in such a way that the transformer must be disconnected from the system before the switch can be operated. The tappings are preferably placed on the winding corresponding to the voltage most likely to vary, usually the high voltage winding. As this winding has the higher number of turns and lower current, the design and the construction details are more easily resolved.

This arrangement enables the transformer to operate at practically constant flux density or to remain within the ±5 per cent limits required by the standards, and this corresponds to an optimum size for the transformer.

The regulation is usually symmetrical above and below the rated voltage, ±2.5 per cent or ±5 per cent, to suit the local conditions. Depending on the way in which the windings have been constructed, the tappings are placed either at the middle of the winding at a break (windings as coils or discs) as in Figure 2.5, or on the outside layer (windings in layers). The connections are brought to a three-pole switch placed in the tank, and operated from the outside on the cover by a handwheel, or for operation at ground level by a handle and operating mechanism.

Two problems arise in the design of tapped windings. Firstly, the winding in the tapping zone, the connections and the switch must be designed to withstand both normal power frequency voltages and also transient voltages, in particular those resulting from reflections at the open ends of the windings. Secondly, switching out portions of windings can alter the balance of the ampere-turns between the primary and secondary windings. The windings must therefore be designed to reduce the short-circuit axial forces to the lowest possible value for all tap positions. A satisfactory solution can be obtained by using a computer program.

These problems become more acute as the regulation range increases. When the regulation exceeds ±5 per cent, the number of positions for regulation tends to increase, the arrangement of windings and tappings becomes more complicated, and this can lead to tapping windings having to be separate from the principal windings. Such complications cause additional cost. It is necessary therefore to limit the range of regulation and the number of tappings to that which is strictly necessary.

TRANSFORMERS REGULATED ON LOAD

On-load regulation avoids interruption to supply due to the transformer being switched out, and allows continuous adjustment of the transformer ratio to be made. The passage from one tap to another, without interrupting the principal circuit, requires the two tappings to be connected together for a very short period. To avoid a complete short circuit, this connection is made through a resistance, or less commonly an inductance.

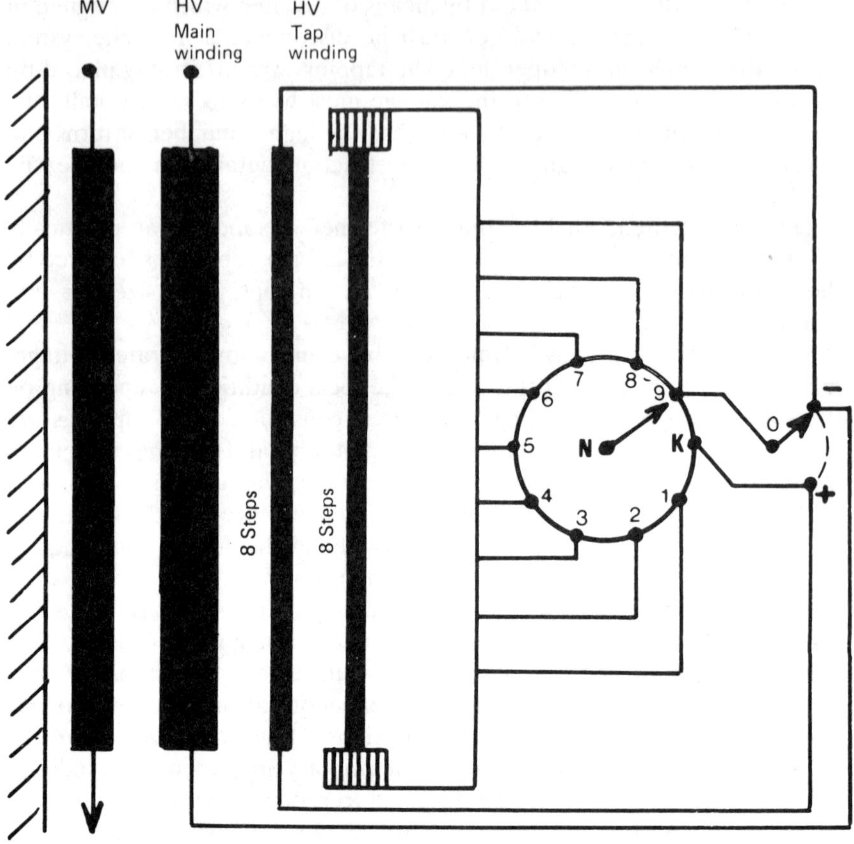

Figure 2.6a. Regulation using coarse and fine windings with 17 operating positions

On-load regulation requires for each phase, therefore, one or more windings for regulation, a tap selector, a resistor and some switches to enable the change to be made.

For transformers connected in star, the taps are always placed at neutral where the dielectric stresses, both power frequency and transient, are lowest. For economy and to reduce size, the tap changing equipment for the three phases can be grouped in a single unit. The construction of the transformer windings and the tap leads is thus simplified.

Transformers connected in delta, and most auto-transformers, require a separate tap changer for each phase generally with common drive. This arrangement inevitably leads to more severe stresses.

Tapping windings may consist of a single winding with n sections and reversing switch giving $(2n+1)$ or $(2n-1)$ positions depending on the

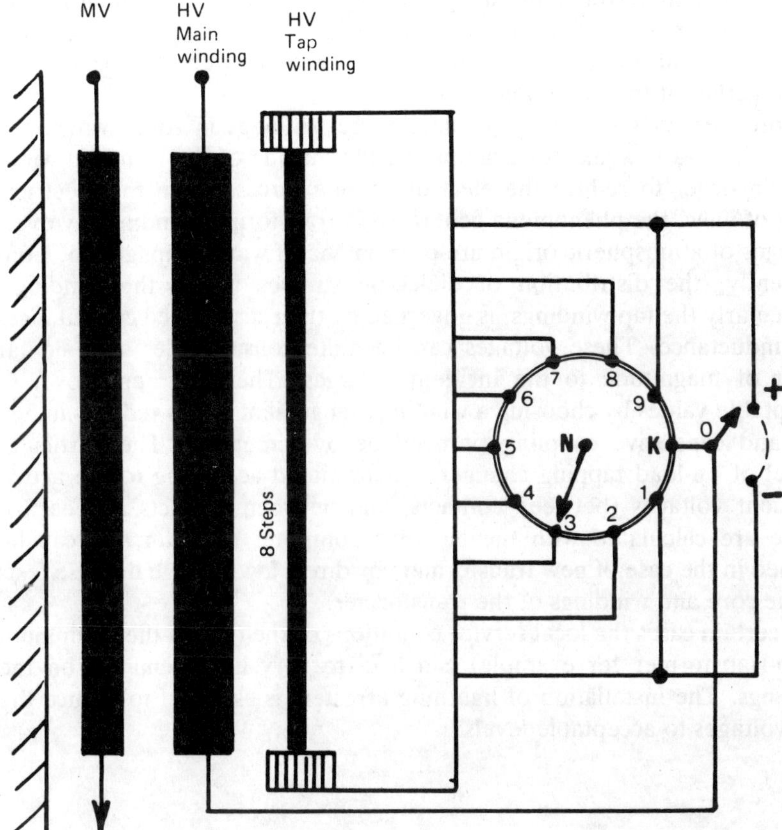

Figure 2.6b. Regulation using reversing switch, with 17 operating positions

construction of the tap changer, or a composite winding formed of a 'coarse' winding comprising a number of turns corresponding to n or $n+1$ steps depending on the type of tap changer, and a 'fine' winding comprising n steps. See Figure 2.6.

The assembly will be able to give $(2n+1)$ or $(2n-1)$ service positions depending on the type of tap changer. In practice, the range of possibilities is such that currently tap changers for 9 to 35 tapping positions can be constructed. The arrangement of the windings in practice is as follows, starting from the core: MV–HV main–HV taps. This gives the best arrangement for the insulation and a simple disposition of the connections.

The electrical and electrodynamic problems merit a special mention. The axial forces generated during a short circuit require one or more separate windings, due to the amount of regulation which can be 20–30 per cent of the nominal voltage. These windings are concentric with the principal windings and are made in such a way as to have the best possible balance of ampere-turns for all tap positions. They comprise windings with equal steps on each side of the middle point of the tapping range, and single layer coils for each tap section, all layers being wound concentrically and together at the same time.

Moreover, it is necessary to place these windings in zones where the magnetic field is weak, for example on the outside of the principal windings, in order to reduce the electromagnetic forces. From the electrical point of view, the phenomena generated in transformer windings by overvoltages of atmospheric origin are phenomena of wave propagation. Consequently, the distribution of dielectric stresses within the windings, particularly the tap windings, is governed by their distributed capacitances and inductance. These voltages can be quite considerable, of a similar order of magnitude to the incident voltages. They are kept down to acceptable values by choosing a winding arangement which reduces inductive and capacitive coupling (if need be by screening). The particular model of on-load tapping changer is determined according to the actual transient voltages (between contacts, and between contacts and earth). These are calculated with the aid of a computer program and can be verified in the case of new transformers by direct low-voltage impulse tests on the core and windings of the transformer.

In certain cases the local service conditions or the type of the equipment (auto-transformer for example) can lead to very high demands on the windings. The installation of lightning arresters is essential to reduce the overvoltages to acceptable levels.

Group 10: Tap changer head
11. Transformer tank cover
12. Tap changer mounting
13. Top cover
14. Top position indicator
15. Connecting pipework
16. Upper gear train
17. Air release plug

Group 20: Oil compartment change-over
    switch
21. Insulating cylinder
22. Upper flange with gasket
23. Lower flange with gasket
24. Sliding contact
25. Terminal

Group 30: Change-over switch assembly
31. Lower gear train
32. Support for 31
33. Supporting rod
34. Drive shaft
35. Cam
36. Energy accumulator with ratchet

36a. Ratchet
37. System of fixed contacts
38. System of moving contacts
39. Current limiting resistors

Group 40: Fine tap selector
41. Selector gear train
42. Coupling
43. Terminal
44. Laminated paper rod
45. Central column of selector with
    collector ring
45a. Collector ring
46. Selector contact arm
47. Selector support
47a. Upper support
47b. Lower support

Group 50: Preselector
51. Preselector drive
52. Laminated paper rod
53. Terminal
54. Preselector moving contact

ON-LOAD TAP CHANGERS

The transformer-mounted on-load tap changer, with rapid change-over using resistor, is commonly used. Results from 50 years of constant progress in both performance and reliability have made it suitable for ever increasing voltages and currents. The current ratings range from 300 to 2000 A and the voltages from 30 to 225 kV. Several types of tapping selector exist for the same rated voltage, to take account of the transient

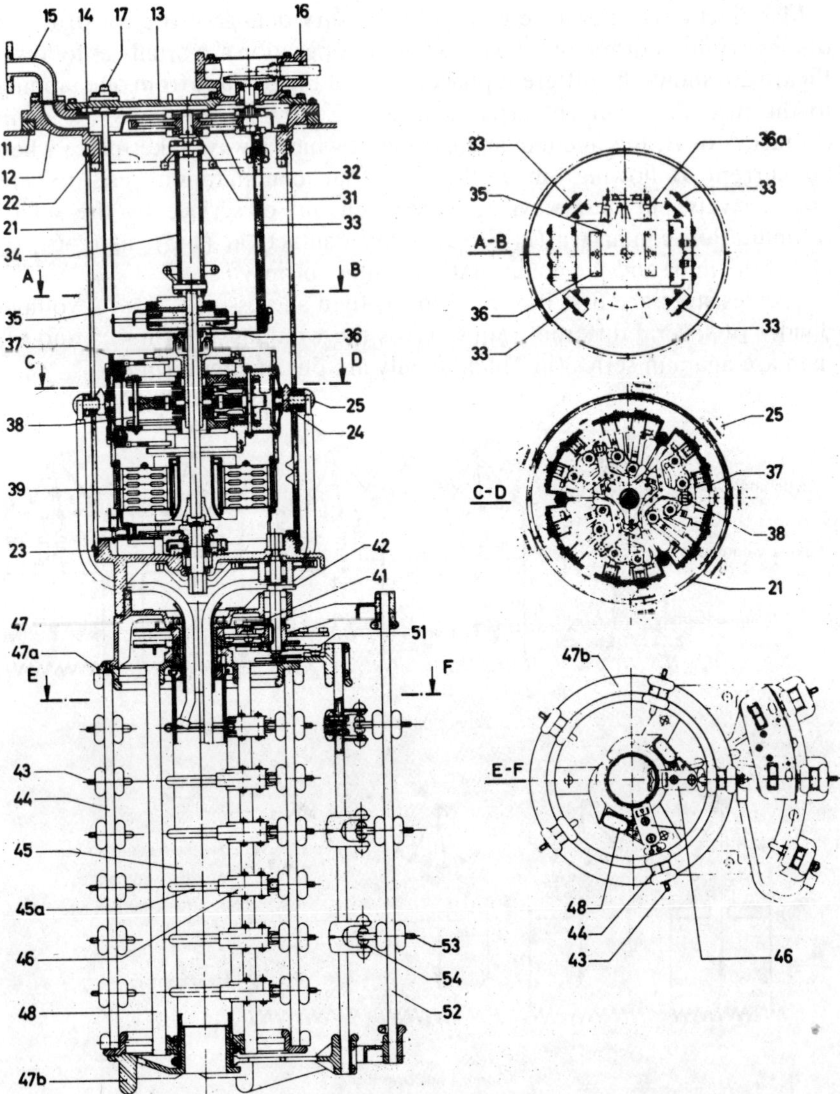

Figure 2.7. On-load tap changer type M, model M III 500-110C-10 19 1 G

voltages in the tapping winding. Two types of regulation are possible, reversing or coarse and fine. The former is suited to medium-power (5–40 MVA) transformers and the latter to large transformers and very high voltage. The commutator consists of an insulating cylinder containing diverter switch, resistors and mechanical transmission, at the lower end of which are found the tapping selector and change-over switch. The whole assembly is immersed in the transformer tank. The oil in the cylinder is separated from that of the tank (Figure 2.7), eliminating the risk of contamination of the transformer oil.

Manufacturers' literature gives all necessary details of the construction of these commutators, but the principles of operation are briefly as follows. Figure 2.8 shows the different phases, 1–9, of the passage from one tapping to the next. The current arriving from the winding leaves by the main contact A or B, the opening and closing of which always takes place when no current is flowing (in 1–9). The four commutation resistors are progressively put into service, then taken out of service by the set of commutation contacts a (in 3), auxiliary contacts a_1 (1–5), a_2 (2–6), b_2 (4–9), b_1 (in 6) and the commutation contact b (in 8).

The resistors are first in series (in 3), then successively form a voltage divider of several different ratios across the winding step (in 4, 5 and 6), then are again in series (in 7) and finally are out of circuit (in 8).

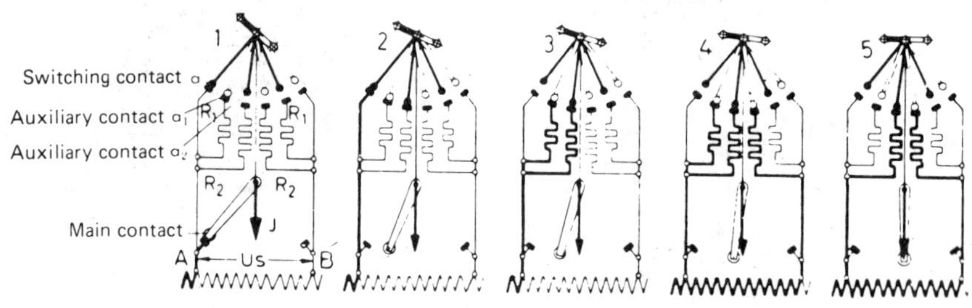

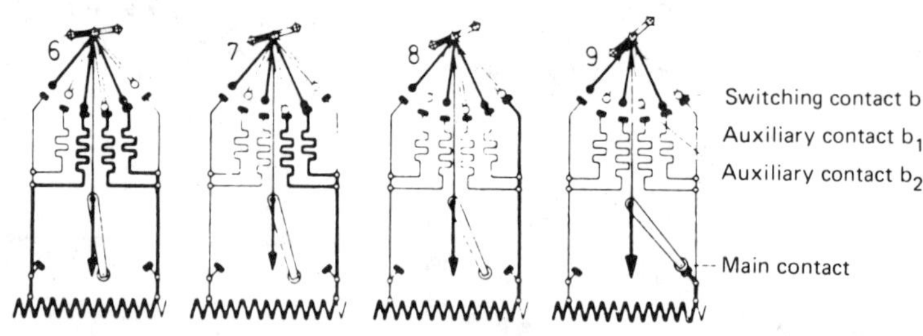

Figure 2.8. Switching procedure for on-load tap changer

The voltage changes gradually during the transition, allowing the use of switches which have to break the in-phase currents at recovery voltages lower than the step voltage.

AUTOMATIC CONTROL OF VOLTAGE

The optimum use of on-load regulation is obtained when a voltage regulator is operated automatically. This voltage regulator has two adjustable parameters: a variable sensitivity from ±0.8 to ±5 per cent of the reference voltage (a range from 1.6 to 10 per cent), and a time delay of 10–200 seconds. With the step voltage of the tapping winding of the transformer these two parameters affect the quality and stability of the regulation.

As long as the voltage remains in the sensitivity range centred on the reference voltage, the regulator does not act. When the voltage falls outside the range, the timing system is started, and if the voltage difference still exists at the end of a fixed time period, the regulator signals the tap changer to change one step (Figure 2.9).

Two conditions must be met. The required sensitivity of control must be obtained, and the number of operations of the tap changer must not be unnecessarily high, to avoid premature wear.

The required sensitivity of control is obtained by choosing the appropriate range and step voltage. Investigations by Electricité de France show

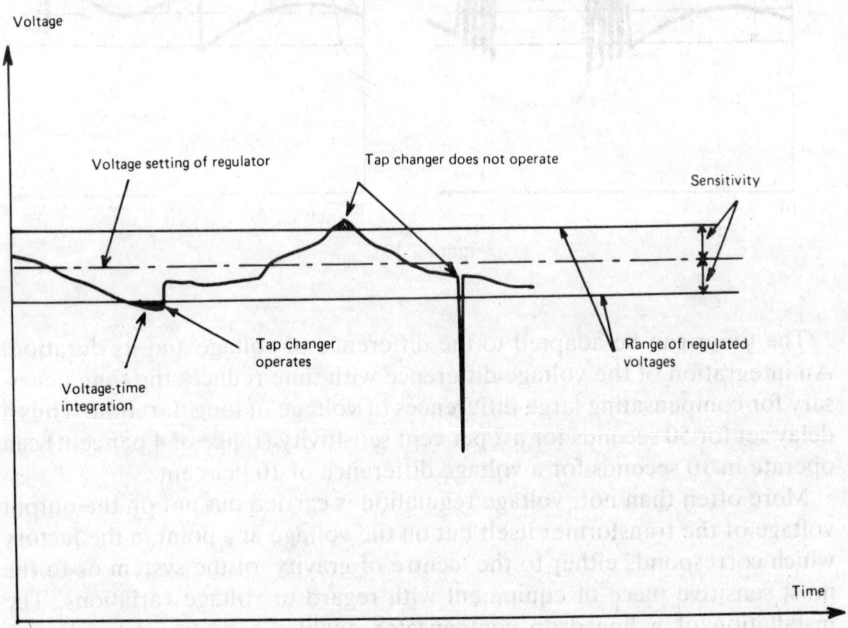

Figure 2.9.

that a step voltage of 1.5 per cent is suitable for distribution systems (with a total of ±12 per cent). For industrial systems this value or even a higher value should, as a general rule, be satisfactory unless there are special local conditions (such as equipment highly sensitive to voltage variations).

To avoid the tap changer operating too often it is better in the first instance for the sensitivity range of the regulator to be from 50 to 100 per cent higher than the step voltage of the tapping winding. A hunting effect results from a ratio for the sensitivity range to step voltage of less than unity (Figure 2.10a). When the ratio is greater than unity a minimum number of regulations recentre the voltage (Figure 2.10b). Moreover, the regulator must not sense voltage variations lasting more than a few seconds; it is with such variations that the time-delay prevents any operation. This becomes even more necessary as the range of sensitivity of the regulator becomes narrower. A ratio of 10 is possible between the number of tap changes for the two extreme values of the timer.

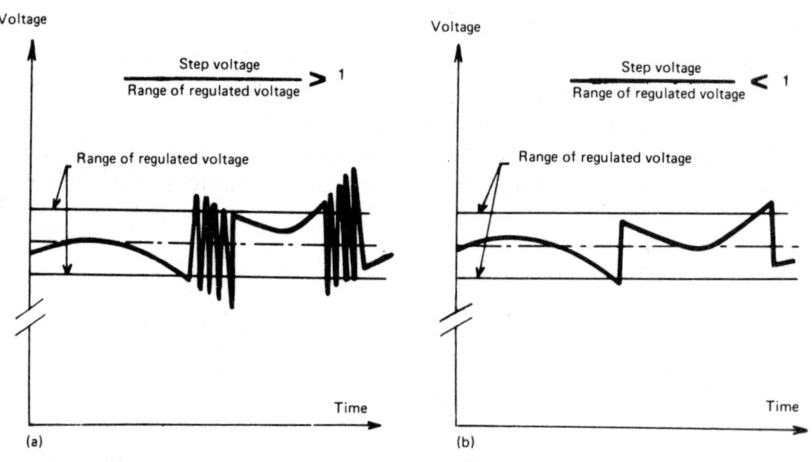

Figure 2.10

The timer can be adapted to the difference of voltage and its duration. An integration of the voltage difference with time reduces the time necessary for compensating large differences of voltage of long duration. Thus a delay set for 50 seconds for a 2 per cent sensitivity (range of 4 per cent) can operate in 10 seconds for a voltage difference of 10 per cent.

More often than not, voltage regulation is carried out not on the output voltage of the transformer itself but on the voltage at a point in the factory which corresponds either to the 'centre of gravity' of the system or to the most sensitive piece of equipment with regard to voltage variations. The installation of a line drop compensator enables resistive and inductive voltage drops up to that point to be simulated. These voltage drops are taken into account by the regulator and added vectorially to its supply voltage to give an image of the actual voltage at the chosen point.

When several on-load regulated transformers supply in parallel, their behaviour under the action of individual regulators would be unstable. With parallel operation therefore it is necessary to provide some equipment to maintain load equilibrium, either by ensuring that the various tap changers (assumed to be identical) step together using a single regulator, or by regulating the current circulating between the transformers to a minimum value by providing additional information for each of the individual regulators. This problem is dealt with in greater detail in Chapter 4, Parallel Operation.

3 Capitalization of losses

THE OPTIMUM EQUIPMENT

When a transformer station is first proposed for an industrial concern, the user will specify some required transformer characteristics based on the expected service conditions. These will include power required, primary and secondary voltages, voltage regulation, winding connections, and method of cooling.

From this specification, the manufacturer of the transformer is able to fix a large number of the possible alternatives, each having a different weight, size, cost and losses.

The minimum cost is obtained by reducing the quantity of active material (core plate, copper), with increased losses, in accordance, of course, with the limitations of temperature rises laid down by the standards. On the other hand, low losses are obtained by increasing the active material in the transformer, so putting up the cost. The cooling equipment is also a factor in both cases, as its cost rises with the amount of losses that have to be dissipated.

The cost of transport, civil engineering works and commissioning have to be added to the ex-works cost of the transformer to obtain the total capital investment. The method of cooling the transformer and the precautions that need to be taken against the risk of combustion of the dielectric affect the amount of civil engineering works, and this can be an element in the choice between several possible solutions.

The purchaser, in collaboration with the manufacturer, must determine the technical/economic optimum, taking into account both the capital investment and the operating costs.

Figure 3.1. Three-phase transformer 20 MVA, 63/20 kV

OPERATING COSTS

Operating costs have fixed and variable components.
  Fixed costs comprise:
  the no-load losses of the transformer, located in the magnetic circuit;
  losses due to cooling, if this consists of only a single stage in continuous operation;
  maintenance costs;
  administrative costs such as insurance premiums.
  Variable costs comprise:
  losses due to the load, varying as the square of the power supplied; these are located in the copper windings;
  losses due to the cooling equipment, pumps, fans; these are included in the variable losses when cooling is carried out in several stages.
When several variants of the same transformer are compared, only the no-load and on-load losses are likely to change and are considered.

From the transformer load diagram and the price of electrical energy, the user can determine the annual cost of the no-load and on-load losses. This cost can be considered as the assured annuity for the estimated life of the transformer obtained from a capital sum invested at compound interest at the time of purchase. This sum is what is meant by the capitalized value of losses.

TOTAL CAPITALIZED COST

The total capitalized cost is therefore the sum of two terms different in nature: the cost of initial installation, and the capitalized cost of losses. As, for any given transformer, these two terms vary inversely with respect to each other, an economic optimum exists for the total capitalized cost resulting from technical and economic considerations.

Theoretically it is possible to carry out a detailed calculation, year by year, by evaluating the variation of the different parameters with time: price of energy, annual losses, rate of interest. The difficulty of forecasting the evolution of these parameters over a 20 year period means that in practice the accuracy is illusory. Constant values are therefore assumed and the calculation is reduced to determining two terms expressing the capital per kilowatt of guaranteed losses: one for the no-load losses, the other for the load losses.

PRODUCER AND USER VIEWPOINTS

The method of capitalizing losses has been the subject of numerous studies. It has been applied for many years, in a general way, to the production and distribution of electrical energy. The points of view are not necessarily the same for the producer of electricity and the industrial user. This divergence can even appear between various categories of user.

The producer of electricity sells one form of energy. The power lost in heat in transformers constitutes for him a permanent loss of income, hence an operating cost. Moreover he has to consider the increased size of equipment (turbo alternators, transformers, instruments, lines) necessary to make up and transport this lost power.

The industrial consumer however uses different types of energy (thermal, mechanical, electrical, human). The electrical losses constitute a permanent cost which must be capitalized. Their influence on the final cost of the product depends on the type of industry. The electrochemical industry, for example, uses electrical energy almost exclusively and this forms an important part of the cost of the products. The capitalization of losses must obviously be taken into account. The metallurgical industry, however, uses a variety of energy forms, among which electrical energy has the highest efficiency. The capitalization of electrical losses, therefore, is not so important.

The initial cost of the installation is, for both producers and users, an investment cost, whereas the other costs are operating costs which influence the service life of the installation. This is an economic consideration and is not necessarily the same as the actual life of the installation which is a technical consideration.

The calculation of the capitalization of losses requires a definition of the capitalized value, the electrical energy tariffs, and the quantity of energy lost annually.

CAPITALIZATION FACTOR

Capitalization reduces all the ongoing expenses during the service life of the transformer to a corresponding capital sum at the time of purchase.

To generate a sum of 1FF per annum over a period of n years, with a rate of interest of r per cent it is necessary to have a capital sum C given by

$$C = \frac{1}{r} - \frac{1}{r(1+r)^n}$$

The values of C for service lives of 5–25 years, at rates of interest of 4–14 per cent, are:

			$r(\%)$			
n	4	6	8	10	12	14
5	4.45	4.21	3.99	3.79	3.60	3.43
10	8.11	7.36	6.71	6.14	5.65	5.22
15	11.12	9.71	8.56	7.61	6.81	6.14
20	13.59	11.47	9.82	8.51	7.47	6.62
25	15.62	12.78	10.67	9.08	7.84	6.87

Thus for a service life of 20 years, at constant losses and rate of interest of 8 per cent, the required capital is 9.82FF per 1FF, i.e. the annual cost of losses must be multiplied by a factor of 9.82 to obtain the capitalized cost.

At the end of each year, for a period of five years, 1FF can be obtained with a starting capital of 3.79FF. If 3.79FF is saved on the purchase price, 1FF can be spent each year for 5 years.

ENERGY TARIFFS (French example)

Electrical energy tariffs are dependent on generating costs, the power demanded, the time at which it is demanded, and the conditions of supply.

This adjustment appears in tariffs applied in France to clients supplied at high voltage. It comprises a fixed charge depending on the power subscribed, and a variable energy price, depending on the voltage, the region and the period of the year.

Two periods apply: *winter* runs from the beginning of October to the end of March, *summer* from the first of April to the end of September. The day is divided, as shown in Figure 3.2, into peak hours, normal hours, and off-peak hours. Sunday is considered as being entirely off-peak. There are five categories of tariffs: P_1 = peak hours, P_2 = normal winter hours, P_3 = off-peak winter hours, P_4 = normal summer hours, P_5 = off-peak summer hours. Depending on the method and duration of use of electrical energy annually, several weighted formulae are possible. One formula often used employs base tariff (long periods of use) for 60 per cent of the maximum power, plus standby tariff (short periods of use) from 60 to 100 per cent of the maximum power.

Figure 3.2 gives the cost per kilowatt hour and the fixed premium for each hour, the winter and summer seasons and for peak tariffs (long period use above 3500 h/annum) and standby tariffs (short period use).

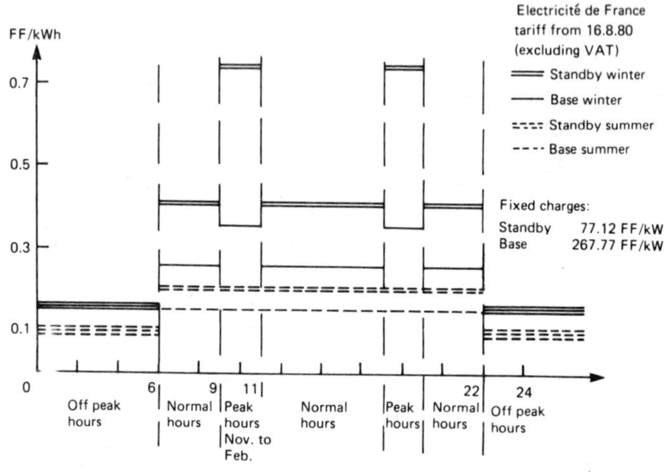

Figure 3.2.

The load factor is the total annual consumption divided by the power supplied. It varies considerably from industry to industry as shown below.

Industrial sector	Use of the maximum load annually	Load factor
Electric steel works	4200 h	0.55
Household equipment	4020 h	0.72
Automobile	3590 h	0.59
Brewery	4420 h	0.66
Footwear	2640 h	0.60
Cement	5600 h	0.76
Confectionery	4700 h	0.62
Cotton	3800 h	0.71
Wool	2500–5300 h	—
Furniture	2610 h	0.51
Paper	5400 h	0.71
Foodstuffs	5290 h	0.71
Tyres	4770 h	0.42
Dairy products	4570 h	0.71
Radio, television	3940 h	0.66
Glass	5820 h	0.49

CAPITALIZATION OF NO-LOAD LOSSES

With the transformer voltage varying very little, the no-load losses are constant. Generally, a transformer permanently on voltage is used annually for about 8000 h and the energy spent is 8000 kWh per kW of loss. Thus, with an interest rate of 14 per cent, a service life of 20 years and an average price of 0.22 FF/kWh the capitalized losses would be

$$8000 \times 0.22 \times 6.62 = 11600 \text{ FF/kW}.$$

Actual cost of losses (k_{Fe}) = number of hours × price/kWh × capitalization factor.

CAPITALIZATION OF ON-LOAD LOSSES

It is advisable, first of all, to estimate the daily variation of the load for different periods of the year. This is best done from experience. Current practice in factories is to carry out regular measurements in the various workshops by inserting recording wattmeters and ammeters in the supply. This information provides a base for estimating requirements for new installations.

Starting with the annual load diagram for the transformer and the costs of energy at any particular time it is possible to calculate the annual cost of on-load losses. The same cost could be obtained by operating at a certain constant power. The use of the transformer is represented by the 'equivalent economic load'. In general it is about 50 per cent of the maximum power, but it can be as high as 100 per cent in factories working continuously. A more approximate calculation can be carried out on the basis of the estimated annual consumption and the average cost per kilowatt-hour. Depending on whether it is on constant load or increasing load with time, the rated transformer power will be equal to or greater than the initial maximum power.

Using the same hypotheses as previously (service life 20 years, interest rate 14 per cent, price 0.22 FF per kilowatt-hour) for a transformer of equivalent economic load of 0.6 times the peak power, which is itself equal to the nominal power, the capitalization of losses due to the load reaches a value (k_{Cu}):

$$8000 \times 0.22 \times 6.62 \times (0.6)^2 = 4200 \text{ FF/kW}.$$

OPTIMUM TOTAL COST

A practical example demonstrates the method and the relative importance of the different factors. Consider a transformer for an industrial use with the following characteristics:

power: 20 MVA;
voltage: 63 kV ± 12%/5.5 kV;
cooling: ONAN (natural circulation of air and oil).

To study the effect of the variation of the interest rate (or price per kilowatt-hour or service life) and annual usage, two groups of values for the capitalization of losses are taken (values in 1974).

Annual Usage	k_{Fe}	$k_{Cu} > 5500$ hours	$k_{Cu} > 3500$ hours
Interest Rate			
7.5%	6000 FF/kW	3000 FF/kW	1500 FF/kW
12.0%	4000 FF/kW	2000 FF/kW	1000 FF/kW

The highest values of capitalization of losses correspond either to a low interest rate or to a high energy cost. The interest rate of 7.5 per cent corresponds to the average value over the last 30 years.

With the aid of a computer programme a large number of variants with no-load losses and variable on-load losses are determined. For each case, the calculation for capitalization of losses is made using the formula above to show the evolution of total cost.

In practice, the rates of capitalization are fixed at the start, and the programme automatically seeks the solution of minimum total cost compatible with the conditions imposed.

It should be noted that technical requirements (maximum flux density in the core, temperature rise of windings during normal operation and on short circuit) set a maximum limit to the losses for a given power. By over-sizing a transformer, however, it is technically possible to reduce the losses to as little as required, but at the expense of the size and cost.

The optimum solutions for the four groups of value of capitalization, where the price of one of the variants has been taken as base, are:

	Capitalization of Losses (FF/kW)			
	6000/3000	6000/1500	4000/2000	4000/1000
Purchase price	119%	100%	102%	87%
Capitalization of no-load losses	26%	26%	21%	21%
Capitalization of on-load losses	63%	45%	52%	37%
Total capitalized price	208%	171%	175%	145%

An examination of the results and details of the calculations reveals some interesting facts.

The minima are relatively flat. A variation of 10 per cent in the purchase price around the minimum causes a 1 per cent increase in the total capitalized price. For larger differences the increases are more rapid.

The capitalized value of the losses is of the same order as the transformer purchase price.

High rates of capitalization lead to low loss, high cost transformers.

The annual use of the transformer, having an influence on the rate of capitalization of losses due to the load, has a marked effect. By reducing from 5500 h/annum (k_{Cu} = 3000 FF/kW) to 3550 h/annum (k_{Cu} = 1500 FF/kW) the losses due to the load increase by more than 40 per cent.

For the same load, the reduction by a third in the rate of capitalization by varying the interest rate and the service life, causes an increase in losses of more than 20 per cent.

This is a particular example and the values which correspond to it could never be generalized. Nevertheless they indicate the trend in the variations and the relative influence of the different factors.

In some practical cases, this method provides the user with the information that enables him to take a decision. For example:

When an extension of an existing installation is being considered, is it more profitable to replace an old transformer by a more powerful new transformer, or to add another transformer?

Taking into account technical progress, when is it less economic to retain an existing transformer than to replace it with a modern one?

In both cases the residual value of the existing equipment and the cost of replacement must be included in the capitalization. The chosen solution will be that of lowest total capitalized cost.

CONCLUSION

The capitalization of losses is a factor to be considered when choosing a transformer. Very precise calculations are not justified, because the accuracy of the results is of the same order as the assumptions made regarding the future evolution of the parameters, bearing in mind also that the minima are not well defined.

Within the limits of a precise, unchanging specification, allowing variation of losses, the factors to be considered, related either to the type of industry and its rate of growth or to economic conditions, are: the cost element of electrical losses in the final cost of the product; load diagram; economic service life; capitalization factor; cost of energy; cost of maintenance; administrative costs; civil engineering costs.

4 Parallel operation

OPERATING CONDITIONS

Two or more transformers operate in parallel when they are connected to the same system on both the primary and the secondary sides.

When the transformers are placed side by side, as in a transformer substation, they are operating directly in parallel. The external impedances of the connecting bars are negligible in comparison with those of the transformers and can be ignored. However, where the parallel connection of the transformers in a system is by overhead or underground cables, these impedances are high and cannot be ignored. In the latter case the load is taken from several points in the secondary system. The distribution of load between the transformers is calculated as for a network.

As it is the more common, the first case only will be considered.

The need to make transformers operate in parallel arises either from the increase in demand for electrical energy resulting from the growth of the system supplied, or from considerations of security of supply or from limitations of the short-circuit power. The objective is to obtain a total maximum power as close as possible to the sum of the rated power of the transformers. This is achieved when the following operating conditions are fulfilled:

1. No circulating current exists in the secondary at no-load.
2. The load is shared in proportion to the rated powers.
3. The currents supplied are all in phase.

These are ideal conditions which can never be fully realized in practice, but no serious disadvantage arises as a result.

Figure 4.1. 340 MVA three-phase transformers 15.5, 15.5/405 kV

Publication No. 76 of CEI and Standard UTE C52–100 prescribe the practical operating conditions for transformers in parallel.

Three-phase transformers with balanced loads are considered in this chapter and an indication is given of the behaviour for other couplings and unbalanced loads.

TRANSFORMER DIAGRAM

Assuming that the load is balanced, the calculation for voltage drop can be made for a single-phase arrangement, as the inductances and resistances of the three phases of the transformer for all practical purposes are equal.

Figure 4.2 shows the standard arrangement consisting of an ideal transformer with no losses, and an impedance in the secondary circuit which is that measured during a short-circuit test.

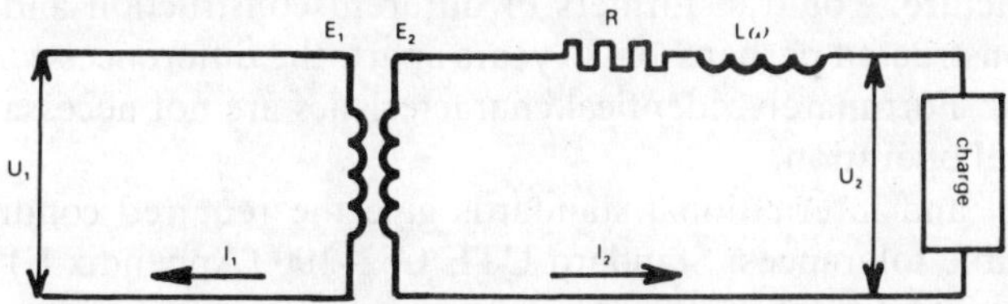

Figure 4.2. An equivalent single-phase transformer circuit

Theoretically, the magnetizing current causes a voltage drop, and the induced primary electromotive force is not equal to the applied voltage. In modern power transformers, the magnetizing current is a fraction of one per cent of the rated current and the corresponding no-load voltage drop is some ten-thousandths of the rated voltage and is therefore negligible.

The KAPP diagram, Figure 4.3, for the transformer secondary, shows the arrangement vectorially. The secondary current I_2 and its phase difference $\emptyset_2$ from the on-load secondary voltage are fixed by the load impedance.

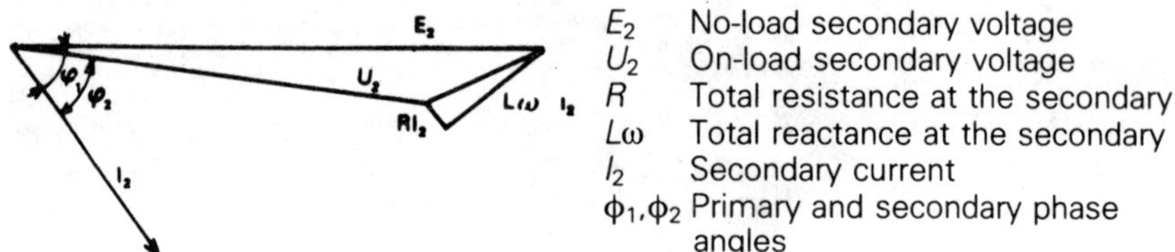

Figure 4.3. KAPP diagram

The resistance of the windings varies with the temperature (4.25 per cent for 10°C), but as only loads approaching full load are of interest, it is the resistance at the reference temperature of 75°C which is taken as the base in standards, guarantees and tests.

To enable the reader to obtain a better appreciation of what actually occurs, the vectorial method is used in the following text rather than analytical methods, although the latter are more accurate. All values are related to the secondary sides, and indices are used to distinguish between different transformers.

REGULATIONS FOR PARALLEL OPERATION

Perfect parallel operation of several transformers is obtained for any total load when the voltage and current diagrams are superimposable. This requires voltages of the same amplitude and phase, currents of equal relative value, and equal resistive and inductive voltage drops.

Even when identical transformers are manufactured together, the last two conditions cannot be exactly met due to the inevitable minor variations in manufacture. For transformers of different construction and different power, constructed perhaps 10–20 years apart, the differences can be quite substantial. Fortunately, identical characteristics are not necessary to permit parallel operation.

National and international standards give the required conditions and the allowable tolerances. Standard UTE C52–100 (Appendix E) specifies:

    (a) compatibility of couplings,

    (b) equal transformation ratios (to ±0.5 per cent),

(c) equal short-circuit voltages (to ±10 per cent),

(d) a guaranteed power ratio between 0.5 and 2.0 and principal tapping.
For clarity in the following text, any load regime is considered as two superimposed operating regimes:

1. No-load operation taking into consideration couplings, transformation ratios and short-circuit voltages.

2. On-load operation on the basis of the resulting no-load secondary voltage, short-circuit voltages and the load of the secondary system.

PARALLEL OPERATION WITH NO LOAD

Firstly, the case of two transformers will be considered. Satisfactory parallel operation requires that the secondary voltages are equal and in phase. Clearly the rated HV and LV voltages of the transformers must be the same.

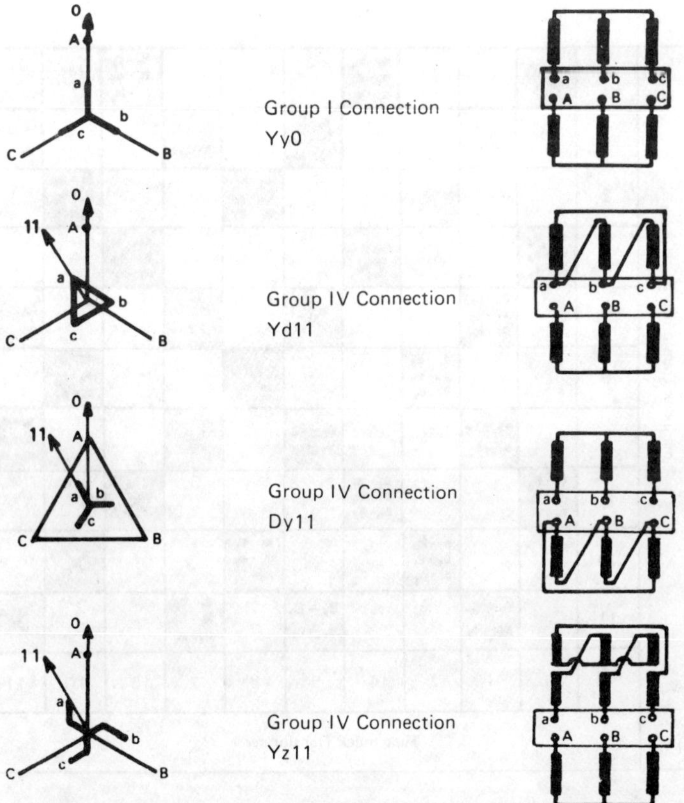

Figure 4.4. Examples of connections and corresponding time indices

Groups of connections

Equality of phase implies that the transformers belong to the same connection group or to compatible groups. The time index which characterizes them corresponds to the position on a dial of a vector representing the voltage at the low-voltage terminal 'a' when the corresponding vector for the voltage at the high-voltage terminal 'A' (left hand terminal when viewing the high-voltage side) is at 12 o'clock (Figure 4.4). The time indices increase in unit steps from 0 to 11 for normal transformers.

There are four groups of connections:

Group I: time indices 0, 4, 8.
Group II: time indices 6, 10, 2.
Group III: time indices 1, 5.
Group IV: time indices 7, 11.

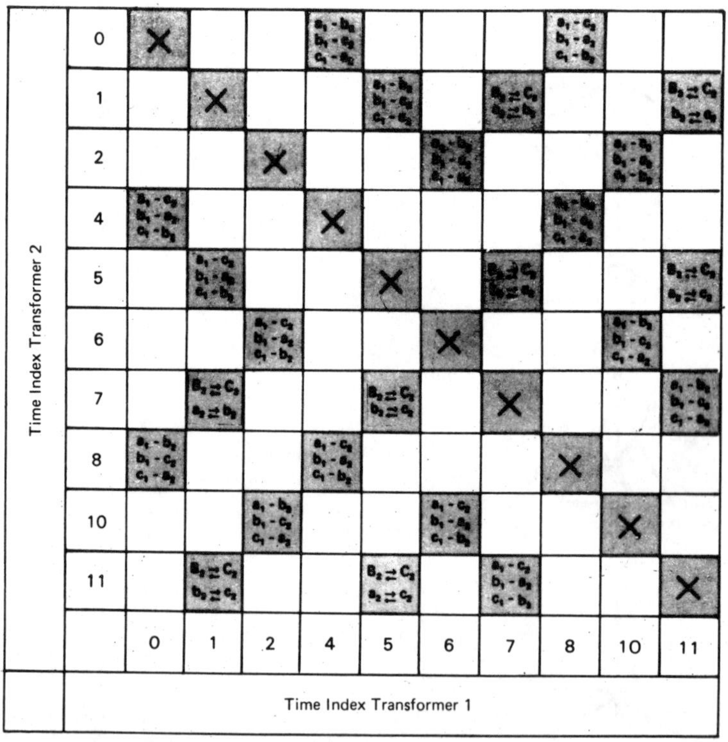

Time Index Transformer 2 \ Transformer 1	0	1	2	4	5	6	7	8	10	11
0	X			$a_1\text{-}b_2$ $b_1\text{-}c_2$ $c_1\text{-}a_2$				$a_1\text{-}c_2$ $b_1\text{-}a_2$ $c_1\text{-}b_2$		
1		X			$a_1\text{-}b_2$ $b_1\text{-}c_2$ $c_1\text{-}a_2$		$B_2\rightleftarrows C_2$ $a_2\rightleftarrows b_2$			$B_2\rightleftarrows C_2$ $b_2\rightleftarrows a_2$
2			X			$a_1\text{-}c_2$ $b_1\text{-}a_2$ $c_1\text{-}b_2$			$a_1\text{-}b_2$ $b_1\text{-}c_2$ $c_1\text{-}a_2$	
4	$a_1\text{-}c_2$ $b_1\text{-}a_2$ $c_1\text{-}b_2$			X				$a_1\text{-}c_2$ $b_1\text{-}a_2$ $c_1\text{-}a_2$		
5		$a_1\text{-}c_2$ $b_1\text{-}a_2$ $c_1\text{-}b_2$			X		$B_2\rightleftarrows C_2$ $b_2\rightleftarrows a_2$			$B_2\rightleftarrows C_2$ $a_2\rightleftarrows c_2$
6			$a_1\text{-}c_2$ $b_1\text{-}a_2$ $c_1\text{-}b_2$			X			$a_1\text{-}b_2$ $b_1\text{-}c_2$ $c_1\text{-}a_2$	
7		$B_2\rightleftarrows C_2$ $a_2\rightleftarrows b_2$			$B_2\rightleftarrows C_2$ $b_2\rightleftarrows a_2$		X			$a_1\text{-}b_2$ $b_1\text{-}a_2$ $c_1\text{-}a_2$
8	$a_1\text{-}b_2$ $b_1\text{-}c_2$ $c_1\text{-}a_2$			$a_1\text{-}c_2$ $b_1\text{-}a_2$ $c_1\text{-}b_2$				X		
10			$a_1\text{-}b_2$ $b_1\text{-}c_2$ $c_1\text{-}a_2$			$a_1\text{-}c_2$ $b_1\text{-}a_2$ $c_1\text{-}b_2$			X	
11		$B_2\rightleftarrows C_2$ $b_2\rightleftarrows c_2$			$B_2\rightleftarrows C_2$ $a_2\rightleftarrows c_2$		$a_1\text{-}c_2$ $b_1\text{-}a_2$ $c_1\text{-}b_2$			X

Time Index Transformer 1

Figure 4.5. Possible parallel coupling of transformers and corresponding connections blank squares. Coupling impossible X. Connection of similar terminals $\rightleftarrows$. Permutation

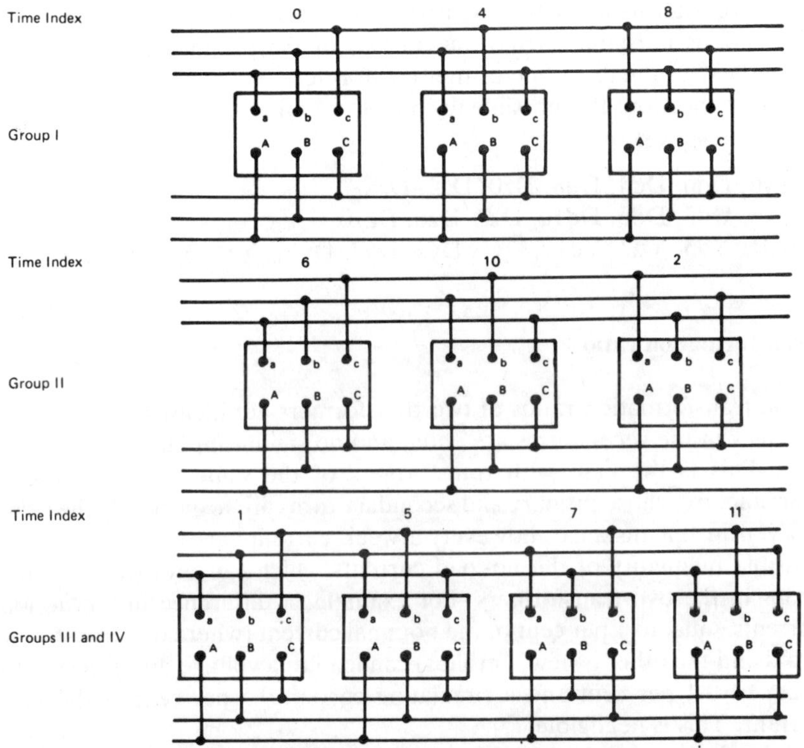

Figure 4.6. Connections of transformers in parallel

The primary and secondary winding connections and the time index are shown on the transformer diagram plate. The most common are as follows:

Group I: Dd0, Yy0, Dz0.
Group II: Dd6, Yy6, Dz6.
Group III: Dy5, Yd5, Yz5.
Group IV: Dy11, Yd11, Yz11.

Figures 4.5 and 4.6 show the possible connections and the schematics for two transformers of any time index.

It is evident that transformers of groups I and II can only be paralleled with transformers of their own group, while transformers of groups III and IV can be interconnected for parallel operation.

Within the same group, transformers with different time indices are connected in parallel by connecting their primaries to corresponding terminals and by phase rotation of the connections to the secondary terminals of one of the transformers.

Where transformers belong to two different groups, interchange of connections must be carried out on two of the primary and two of the secondary terminals of one of the transformers. To summarize, by appropriate connections it is possible to connect together transformers within the following series:

Yy0, Dd0, Dd4, Dd8, Dz0, Dz4, Dz8.
Yy6, Dd2, Dd6, Dd10, Dz2, Dz6, Dz10.
Yd1, Yd5, Yd7, Yd11, Dy1, Dy5, Dy7, Dy11, Yz1, Yz5, Yz7, Yz11.

Transformation ratio

If the transformation ratios of two transformers are identical, the no-load voltages in the secondaries are equal and no significant current will circulate. This is the case with transformers of the same power and same manufacture whose primary and secondary turns are respectively the same.

Even in this instance, however, a weak current can appear due to the possible inequality of the no-load currents which cause different voltage drops in the two transformers. For example, a difference in the no-load currents equal to 1 per cent of the nominal current (where one transformer is old and the other is new, perhaps) can lead to a voltage difference of the order of 0.1 per cent and a circulating current 0.5 per cent of the rated current. This is negligible.

This phenomenon can also occur when connecting two transformers of the same group or of compatible groups having different time indices, because the no-load currents are not identical in the three phases of a transformer.

The effect of a difference in the transformation ratios is more marked. Because of a different construction or different power rating it is generally impossible to obtain the number of primary and secondary turns in precisely the same ratio. For a doubled power, the number of primary and secondary turns are reduced by 30 per cent due to the increase in the section of the core. The no-load voltages of the two transformers are slightly different and a circulation current arises which is limited by the sum of the short-circuit impedances of the two transformers.

$$I = \frac{\Delta u}{E_{cc1} + (P_1/P_2) \times E_{cc2}} \times 100$$

where Δu = voltage difference for no load (in per cent),
    E_{cc1}, E_{cc2} = short-circuit voltages (in per cent),
    P_1, P_2 = rated powers of transformers,
    I = circulation current as percentage of rated current of transformer 1.

The transformer with the lowest power is taken as the base for the calculation.

$$P_1 = 5000 \text{ kVA} \qquad P_2 = 10\,000 \text{ kVA}$$
$$E_{cc1} = 8 \text{ per cent} \qquad E_{cc2} = 10 \text{ per cent}$$
$$\Delta u = 0.5 \text{ per cent}$$
$$I = \frac{0.5 \times 100}{8 + (5/10) \times 10} = 3.8 \text{ per cent of } I_1.$$

The current voltage diagram is shown in Figure 4.7. The current is considerably out of phase with the voltage and remains so whatever the load on the transformer. Although it does not cause high losses (0.15 per cent of full load losses in the preceding case), it is none the less undesirable as it causes voltage drops and limits the maximum power.

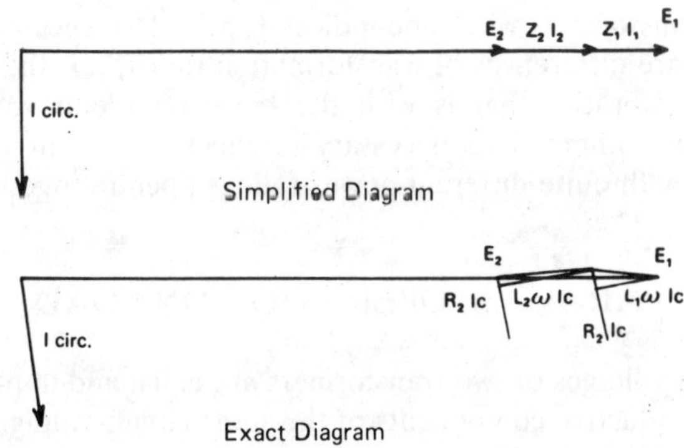

Figure 4.7. No-load circulating current due to different transformation ratios

The formula above, although approximate, is adequate for the general case. The precise formula would be:

$$I = \frac{\Delta u}{\sqrt{(R_1 I_1 + \dfrac{P_1}{P_2} R_2 I_2)^2 + (L_1 \omega I_1 + \dfrac{P_1}{P_2} L_2 \omega I_2)^2}} \times 100$$

where $R_1 I_1$, $R_2 I_2$, $L_1 \omega I_1$, $L_2 \omega I_2$ are the active and reactive components as a percentage of the short-circuit voltages.

This formula simplifies if

$$\frac{L_1 \omega}{R_1} = \frac{L_2 \omega}{R_2},$$

i.e. if the two KAPP triangles are similar.

In practice, the manufacturer adjusts the transformation ratio of the new transformer to a value as close as possible to that of the existing, within an accuracy of 0.1 per cent or better if the voltages are sufficiently high. This degree of accuracy can only be obtained for one operating position. With other tappings the accuracy will generally be a little less. This is why the standards refer to the median position only for a guarantee agreement.

General case

In the case of n transformers having to operate in parallel, the rules made at the beginning of this chapter can be generalized. For example, parallel connections can be made between

(a) three transformers with time indices of 0, 4, 8 respectively,
(b) three transformers with time indices 2, 6, 10 respectively,
(c) four transformers with time indices 1, 5, 7, 11 respectively.

Where there are differences of transformation ratios, it is the transformer with the lowest ratio, that is with the highest no-load voltage, which supplies all the others. It is necessary to check this point when several transformers with quite different power ratings operate together.

PARALLEL OPERATION ON LOAD

If the no-load voltages of two transformers are equal and in phase and the resistive and inductive components of the short-circuit voltages are equal, then the parallel operation is perfect. Together they can supply power equal to the sum of the rated power of each.

In practice, it is not like this because of the type of fabrication, the year of manufacture, changes in specification, etc.

Similar KAPP triangles

Suppose the short-circuit voltages are different but the KAPP triangles are similar. As the effective voltage drop must be the same for both transformers, the load current I is distributed between then inversely as the short-circuit voltages and in proportion to the rated power.

$$I_2 \times \frac{E_{cc2}}{P_2} = I_1 \times \frac{E_{cc1}}{P_1}$$

therefore, $$I_1 = \frac{E_{cc2}}{(P_2/P_1) \times E_{cc1} + E_{cc2}} \times I$$

$$I_2 = \frac{E_{cc1}}{E_{cc1} + E_{cc2} \times (P_1/P_2)} \times I$$

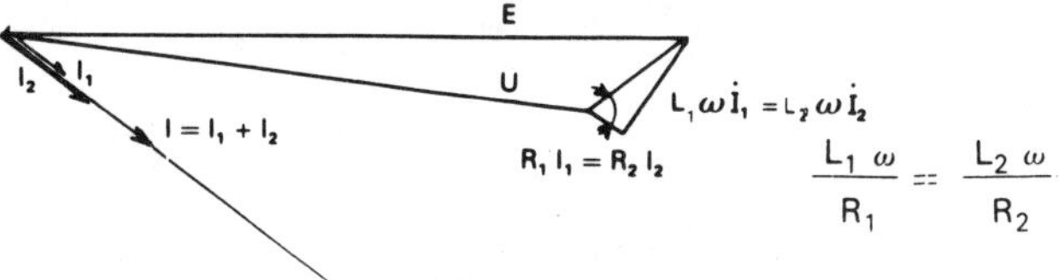

Figure 4.8. Distribution of current between two transformers having similar KAPP diagrams

These two currents are in phase.

As the load increases, the transformer with the lower short-circuit voltage is the first to reach its rated current and this fixes the total current that can be taken.

If $E_{cc1} < E_{cc2}$

$$I_{1n} = \frac{E_{cc2} \times I_{max}}{(P_1/P_2) \times E_{cc1} + E_{cc2}}$$

$$I_2 = \frac{E_{cc1} \times I_{max}}{E_{cc1} + E_{cc2} \times (P_1/P_2)}$$

$$I_{max} = I_{1n}\left(1 + \frac{P_2}{P_1} \times \frac{E_{cc1}}{E_{cc2}}\right)$$

$$P_{max} = P_1\left(1 + \frac{P_2}{P_1} \times \frac{E_{cc1}}{E_{cc2}}\right)$$

$$= P_1 + P_2 \times \frac{E_{cc1}}{E_{cc2}}.$$

The set of curves in Figure 4.9 shows the fraction of the sum of the nominal powers that it is possible to obtain for two transformers with variable ratios P_1/P_2 and E_{cc1}/E_{cc2}, the second ratio being less than or equal to I.

$$\frac{P_{max}}{P_1 + P_2} = \frac{P_1 + P_2(E_{cc1}/E_{cc2})}{P_1 + P_2}$$

$$= 1 - \frac{P_2}{P_1 + P_2}\left(1 - \frac{E_{cc1}}{E_{cc2}}\right)$$

Figure 4.9 shows the limits of guarantee of parallel operation fixed by the UTE standard and the limit below which the maximum power becomes less than the power of the largest transformer.

Where several transformers are in parallel, the calculation for maximum power is carried out using the transformer with the lowest short-circuit voltage as the reference unit. As this transformer reaches its nominal power the remainder will not be fully loaded, and

$$P_{max} = P_1 + P_2\frac{E_{cc1}}{E_{cc2}} + P_3\frac{E_{cc1}}{E_{cc3}} + \dots + P_n\frac{E_{cc1}}{E_{ccn}}$$

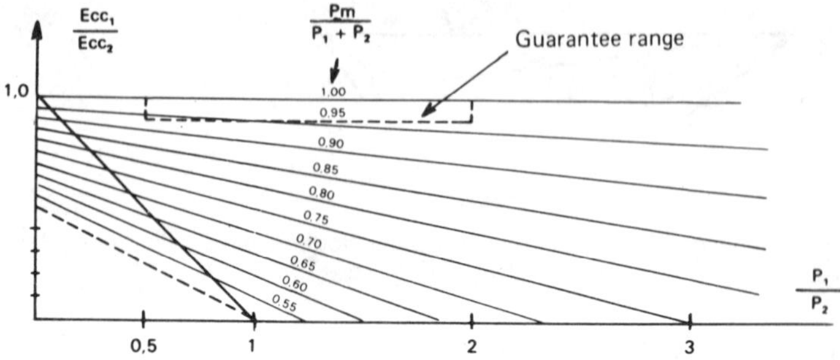

Figure 4.9. Maximum power of two transformers in parallel, as a function of the ratio of rated powers and ratio of short-circuit voltages (short-circuit voltage E_{cci} assumed to be lowest)

From this it can be concluded that it is preferable to have a lower short-circuit voltage for the more powerful transformers. However, the rules for sizing transformers (economic sizing, resistance to short circuit) give an opposite conclusion.

The requirements of parallel operation often lead to a special and non-optimum size, and therefore an additional cost.

Dissimilar KAPP triangles

In practice the KAPP triangles are not similar and the active component RI per cent decreases as the power increases. Figure 4.10 shows the voltage diagram of two transformers. The voltage drop is the same for each transformer but the components are unequal. There is a phase difference between the two currents I_1 and I_2, and the total current is the vectorial sum. The angle formed by the two currents is constant whatever the load and is equal to the difference of the angles α_1 and α_2 of the KAPP triangles. The currents are in inverse ratio to the impedances, as above.

$$I_1 \times \frac{E_{cc2}}{P_2} = I_2 \times \frac{E_{cc1}}{P_1}$$

$$\text{with } E_{cc1} = \sqrt{(R_1 I_1)^2 + (L_1 \omega I_1)^2}$$

$$E_{cc2} = \sqrt{(R_2 I_2)^2 + (L_2 \omega I_2)^2}$$

The phase difference between the currents further reduces the maximum available power. As Figure 4.11 shows, however, the effect is small. For a power ratio of 2 and equal short-circuit voltages, the angle θ is about $5°$, corresponding to a power reduction of less than 0.1 per cent. In practice therefore this correction is ignored, as is the correction for the variation of resistance with heating due to the load, which is of the same order.

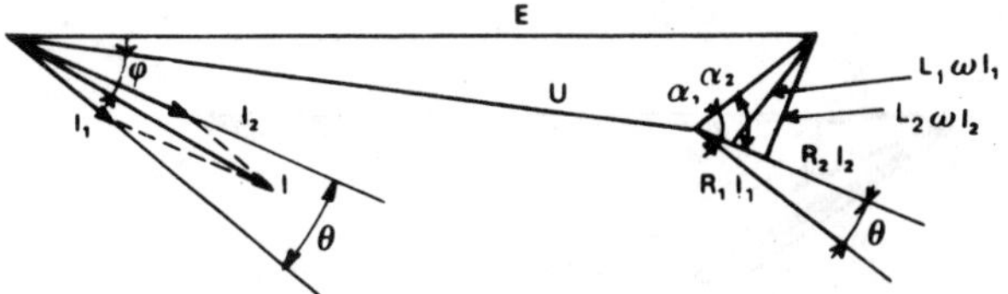

Figure 4.10. Distribution of current between two transformers having dissimilar KAPP diagrams

On-load currents

The actual load currents are obtained by superimposing the no-load circulating current onto the load currents. It has been seen, however, that the manufacturer ensures that the no-load circulating current is as low as possible and its effect is negligible. The load currents can therefore be considered to be the actual currents.

METHODS OF IMPROVING PARALLEL OPERATION

There are cases where it is not possible to obtain equal short-circuit voltages; for example, if the short-circuit voltage of an existing transformer is too low and the new (more powerful) transformer constructed to meet the conditions has too high a short-circuit power on the secondary side. The short-circuit forces can then require over-sizing of the transformer.

The standards provide for two methods of correction, by means of an external reactance or by adjusting the transformation ratio.

External reactance

An external reactance is connected in series to the primary or secondary of the transformer with the lowest short-circuit voltage. If the installation conditions allow it, this solution, although costly, is the best. This is similar to the preceding case.

The power rating of the reactance is given by the formula:

$$P_R = P_1 \times \frac{E_{cc2} - E_{cc1}}{100}$$

where P_1 is the power rating of the transformer with the lowest short-circuit voltage E_{cc1}, and E_{cc2} is the short-circuit voltage of the other transformer. Where there are two transformers, one with $P_1 = 1000$ kVA and $E_{cc1} = 4.5$ per cent, and the other with $P_2 = 2000$ kVA and $E_{cc2} = 6.2$ per cent, the reactance necessary would have power

$$P_R = 1000 \times \frac{6.2 - 4.5}{100} = 17 \text{ kVA}.$$

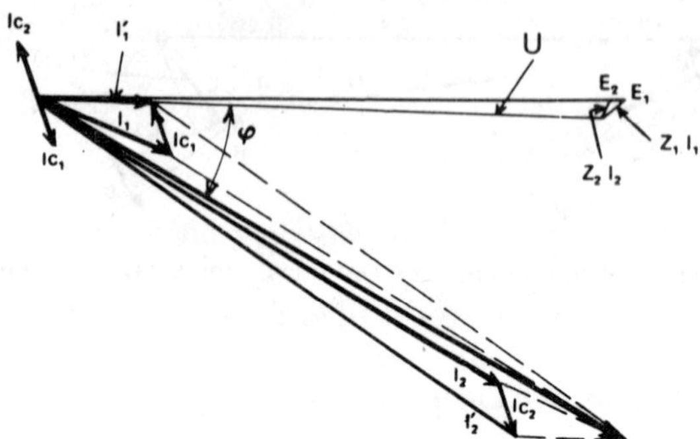

Figure 4.11. Improvement to parallel operation by modifying the transformation ratio I'_1 I'_2 Components of current I before modification $.I_1$ I_2. Components of current I after reduction of transformer 2 voltages from E_1 to E_2 $.I_{c1}$ I_{c2}. No-load circulating currents $.I$. Secondary load current

When the current intensity on the low-voltage side is sufficiently high, e.g. 220 V or 380 V, and the correction is not very large, some gapped magnetic circuits can be placed around the bus bars to form a small simple reactance.

Modification of the transformation ratio

The other possibility is to adjust the transformation ratio of the new transformer to create a circulation current that will reduce the excessive load current, so allowing an increased total maximum load.

The voltage and current diagram is shown in Figure 4.11. It is clear that the circulating current is very much out of phase and so is most effective when the consumption of reactive energy is high, which the consumer must avoid as much as possible.

The effectiveness is further limited if the power ratio is high, because the standards allow a circulation current equal to 15 per cent of the lowest nominal current. At no load the loss due to this current amounts to 2.25 per cent of the load losses for the transformer being considered; this is about 10 per cent of the no-load losses.

PARALLEL OPERATION OF TRANSFORMERS WITH ON-LOAD TAP CHANGERS

If the transformers intended for parallel operation are identical in terms of power, voltage, number and size of steps of reactance voltages, there is no problem. The on-load tap changers are usually provided with electric controls activated manually, by pushbutton or automatically. The automatic system must somehow ensure that all controls obey the commands to change voltage at the same time so keeping all the transformers at the same tap position.

Although a tap difference between two transformers is not harmful, because the regulation steps are quite small, it is important to correct it if it occurs accidentally (for example, because of malfunction or late operation of one of the controls). It is therefore necessary to check that the controls carry out the orders given to them. Several types of equipment of varying sophistication exist for this purpose.

After initially establishing conformity manually, the simplest equipment maintains conformity by not allowing an order to be executed until all the controls have responded to the preceding order. In case of malfunction, the system stops and emits a fault signal.

With a more sophisticated type of equipment, an automatic control only gives a signal to operate if all the regulators are in the same position. The system stops if a control panel does not react normally, a time-delay taking into account the possible difference of speeds of operation. It is even possible to prevent all switching if all the tap changers are not in the same position. Some systems allow automatic realignment of the controls if an accidental out-of-step operation takes place.

This situation arises when the transformers, having different numbers of steps and different voltages per step, must operate in parallel. The solution is an electrical one. The reference voltage is from a voltage transformer connected to the common busbars. The magnitude and phase of the currents is measured for each of the transformers, and from this is deducted the voltage correction, applied to each as a function of the difference from a balanced distribution of currents. The unbalanced transformers receive signals from their regulators to reduce this difference. A certain amount of insensitivity must be introduced to avoid hunting. A check on the response to these signals is carried out by measuring the current circulating between the transformers.

PARTICULAR CASES

Single-phase transformers do not pose any special problems. The conditions for good parallel operation are the same and the secondary voltage must be in phase.

Transformers with three windings can only operate in parallel, terminal to terminal at the primary secondary and tertiary, if they all have the same transformation ratios and the same short-circuit voltages of the windings taken in pairs.

Furthermore, parallel operation of the primary and secondary only, with the tertiaries supplying separate circuits, is possible only if the secondary–tertiary short-circuit voltage is equal to the sum of the primary–secondary and primary–tertiary short-circuit voltages, all referred to the same power. Only under these conditions does the load of the tertiaries have no influence on parallel operation. If this condition is not fulfilled, the tertiaries must remain on no-load. This condition applies equally to two transformers in parallel, where one has three windings and the other has two.

We have assumed throughout that the load is a balanced three-phase one. With loads considerably out of balance, disturbances can occur if the transformers do not have the same HV and LV connections, even though they may have the same time index. For example, with two transformers Dy1 and Yd1 connected in parallel, a single-phase load applied between phase and neutral of the first transformer is not shared by the second.

ECONOMIC PARALLEL OPERATION

When several transformers are likely to be connected in parallel to a distribution system with a variable load, it is possible to vary the number of transformers in service according to the load to reduce the sum of the transformer losses to a minimum.

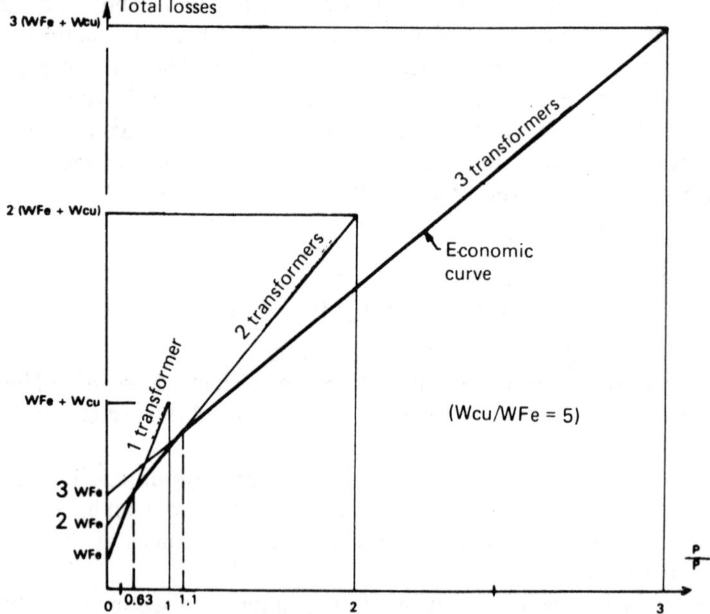

Figure 4.12. Successive switching-in of three identical transformers to obtain minimum losses.

Suppose that there are n identical transformers. They each have power p, no-load losses W_{Fe} and on-load losses W_{cu}. As the load increases from zero, a single transformer is at first sufficient. The losses vary as shown in Figure 4.12, where the horizontal co-ordinates are on a quadratic scale. At a certain point the connection of a second transformer reduces the total losses. The equilibrium point is reached when:

$$W_{Fe} + \left(\frac{p}{P}\right)^2 \times W_{cu} = 2W_{Fe} + 2\left(\frac{p/2}{P}\right)^2 W_{cu}$$

from which $\dfrac{p}{P} = \sqrt{\dfrac{2W_{Fe}}{W_{cu}}} \sim 0.6$ to 0.7.

As the power increases, a third transformer should be connected when:

$$2W_{Fe} + 2\left(\frac{p/2}{P}\right)^2 W_{cu} = 3W_{Fe} + 3\left(\frac{p/3}{P}\right)^2 W_{cu}$$

from which $\dfrac{p}{2P} = \sqrt{\dfrac{3}{2}\dfrac{W_{Fe}}{W_{cu}}} \sim 0.5 \; to \; 0.6.$

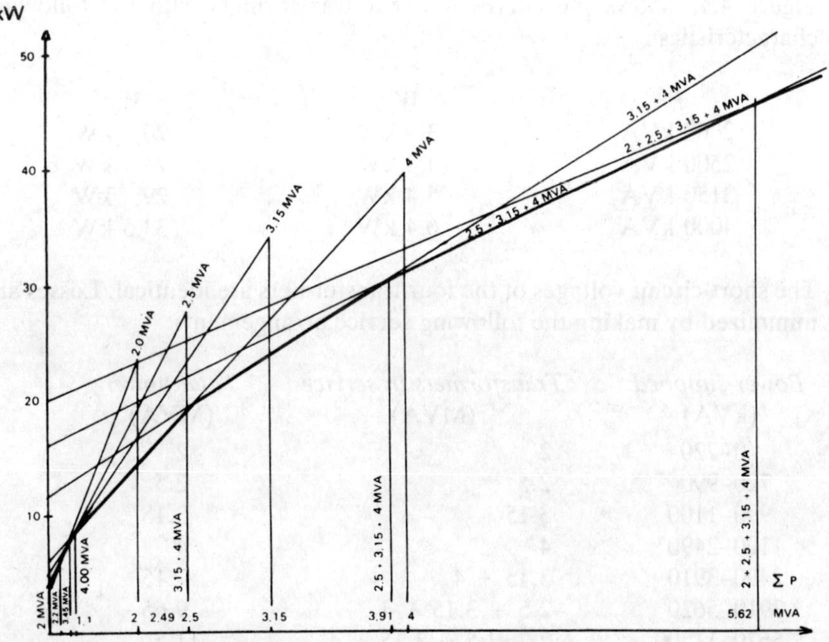

Figure 4.13. Successive switching-in of four transformers of 2, 2.5, 3.15 and 4 MVA to obtain minimum losses

For the $(m+1)^{\text{th}}$ transformer:

$$\frac{p}{mP} = \sqrt{\left(\frac{m+1}{m}\right)\frac{W_{Fe}}{W_{cu}}} \rightarrow 0.4 \text{ to } 0.5.$$

A rough guide is to connect another transformer when those in service are at half load.

As the temperature of the windings varies with the load, the losses are in fact less than those obtained from the standard quadratic equation. The points of equilibrium are therefore slightly below those indicated previously.

When there are several transformers of different power, the equilibrium points can be determined either graphically or by a calculation as shown above. As extreme accuracy is not required, the first method is the simplest. The method is as follows:

1. Determine all the possible combinations of transformers with the maximum power delivered (allowing for possible differences in short-circuit voltages), the total no-load losses and the losses due to the total load for each combination.
2. Draw the curve for the total losses on-load ($W_{Fe} + W_{cu}$) against the square of the total power supplied.
3. Determine the polygon giving the minimum losses for each power.

Figure 4.13 shows the curves for four transformers with the following characteristics:

P	W_{Fe}	W_{cu}
2000 kVA	3.9 kW	20 kW
2500 kVA	4.5 kW	24 kW
3150 kVA	5.4 kW	29 kW
4000 kVA	6.4 kW	33.5 kW

The short-circuit voltages of the four transformers are identical. Losses are minimized by making the following service arrangements:

Power supplied (kVA)	Transformers in service (MVA)	Total power (MVA)
0–720	2	2
720–990	2.5	2.5
990–1100	3.15	3.15
1100–2490	4	4
2490–3910	3.15 + 4	7.15
3910–5620	2.5 + 3.15 + 4	9.65
5620–11500	2 + 2.5 + 3.15 + 4	11.5

In fact, using transformers of 2.5 and 3.15 MVA is not particularly helpful, and the increase would be from 2 to 4 MVA under the following conditions:

| 0–930 | 2 | 2 |
| 930–2490 | 4 | 4 |

The saving in losses is given by the difference of ordinates between the straight-line characteristics of the different arrangements. It is obviously out of the question to make constant adjustments to the system to adapt it to the energy demand. The method can only be applied to slow variations of demand occurring daily (off peak hours, peak hours) or seasonal (summer, winter).

The voltage drop, the short-circuit power and the security of service must also be taken into account according to the characteristics of the system.

INFORMATION NEEDED FOR PARALLEL OPERATION

The consumer expecting to install a new transformer to operate in parallel with existing transformers, must supply the manufacturer with the following information, which can be obtained from the rating plates on the existing units:

name of manufacturer,
place of manufacture (factory),
series number,
type of fabrication,
reference standard,
frequency,
number of phases,
nominal power, voltages and currents of the various windings,
number and steps of the tapping range,
winding connections and time index (or connection schematic),
no-load losses,
losses due to the load and the short-circuit voltage at the rated current at the principal tapping and, if necessary, the highest and lowest tappings.

5 Cooling

SOURCE OF LOSSES IN A TRANSFORMER

During its operation a transformer is the source of energy losses, the majority of which are located in two principal parts:

1. The magnetic circuit, where they arise because of the variation of alternating flux in the magnetic core. They are related to the induction and hence the applied voltage.
2. The windings, where they are primarily due to the I^2R losses, and also to the eddy currents. They are related to the current and hence the load.

Losses are also generated in the connections, tap changers and bushings. They can be linked with the losses referred to above, and appear in the same way in some materials of good electrical conductivity.

The leakage flux from the windings, terminals and connections can also create parasitical losses by inducing eddy currents in neighbouring non-active metallic components, such as the fastenings, tank and cover. The manufacturer must reduce these to the absolute minimum.

All these losses cause heating of the corresponding parts of the transformer and some method of cooling must be introduced.

The rated duty of a transformer (i.e. its rated voltage and power) is closely related to its heating because of the restrictions imposed by the insulating materials used, and to a lesser degree by the increase of the I^2R losses with temperature.

Figure 5.1. 30 MVA 5500−5500/63 000 V transformer with natural cooling

The transmission of heat takes place usually in several ways:

(a) conduction from the various parts of the transformer to their surface,
(b) convection in a gaseous dielectric which is both an insulator and carrier of heat (as with dry transformers),
(c) convection in a liquid dielectric which transmits heat to the cooling medium in a heat exchanger (as with immersed transformers).

TEMPERATURE STANDARDS

The admissible values of heating are fixed by publication 76, 'Power Transformers', of the International Electrotechnical Commission (IEC) which is followed by most national standards.

The following are specified: maximum temperature of the cooling medium (air or water), the temperature rise of the windings, the magnetic circuit and other parts, and the liquid dielectric if it is an immersed transformer. These values are shown in Tables 5.1 and 5.2.

Table 5.1 Temperature-rise limits for dry type transformers

Part	Method of cooling	Insulation class*	Maximum temperature-rise limit, °C
Windings (measured by variation of resistance)	Air, natural or forced ventilation	A E B F H	60 75 80 100 125 150**
Cores and other parts (measured by thermometer)			
(a) adjacent to windings	All		(a) same value as for windings
(b) not adjacent to windings			(b) the temperature shall, in no case, reach a value that will damage the core itself, other parts or adjacent materials

* In accordance with IEC Publication 85, Recommendations for the Classification of Materials for the Insulation of Electrical Machinery and Apparatus in Relation to their Thermal Stability in Service.
** Temperatures of 150° and even higher can be accepted for certain materials, after discussion between the manufacturer and purchaser.

The average temperature of the windings can be measured by the variation of electric resistance during a temperature-rise test. The heating of the other parts can be measured by thermometer.

Table 5.2 Temperature-rise limits for transformers immersed in oil

Part	Method of cooling	Oil circulation	Maximum temperature-rise °C
Windings, Insulation Class A (measured by variation of resistance)	Natural By forced ventilation By circulation of water internally	Natural	65
	By forced ventilation By external water cooling	Forced Forced around the windings	65 *
Top Oil (measured by thermometer)			60, when the transformer is equipped with a conservator or sealed. 55, when the transformer is neither equipped with a conservator nor sealed.
Cores and other parts (measured by thermometer placed on the surface)			The temperature must not under any circumstances reach a value likely to damage the magnetic circuit or adjacent parts.

* For transformers with forced circulation of oil around the windings, the heating limit can be increased to 70°C. Verification that the oil circulation is forced around the windings can be made from an examination of the drawings or tests by checking, for example, that the temperature difference between copper and oil is proportional to the losses.

CLASSIFICATION OF INSULATING MATERIALS

Solid insulating materials are divided into seven classes depending on the admissible temperature, as follows:

Class:	Y	A	E	B	F	H	C
Temperature limit (°C):	90	105	120	130	155	180	>180

For transformers immersed in a dielectric oil, the insulation belongs to class A, which as well as natural fibres (cotton, silk, wood) and artificial fibres (cellulose acetate, polyamides, regenerated cellulose), includes paper, press board, layered cloth and film.

Dry transformers can belong to any one of these classes, but the most common are B, F and H.

The materials must have mechanical, electrical and thermochemical characteristics that guarantee their correct behaviour at the maximum temperature and their mutual compatibility.

Ageing of the insulation, which is generally taken as a loss of 50 per cent of the mechanical properties, increases with operating temperature. According to Montsinger's law, the life of an insulator decreases by half for an increase in service temperature of 5–8°C. The temperature and heating limits set by the standards ensure an estimated normal life of 20–25 years. However, even if a transformer is subjected to a constant load, the maximum temperature of the insulation near the hot spot can vary with the ambient temperature. The ageing is therefore strongly accelerated at times but retarded at others, and the average is taken. This aspect of the subject will be examined closely in Chapter 6, Overloads.

FLUID COOLANTS AND INSULATORS

The heat generated by the losses is transmitted to a fluid coolant, air or water, in a heat exchanger. The convection coefficient which represents the ability of a fluid to transmit heat by natural convection depends on five factors: its density, coefficient of thermal expansion, viscosity, specific heat, and thermal conductivity.

Under identical geometric and thermal conditions, oil is 12 times, and water 100 times, more efficient than air. Air has the advantage of being more readily available in unlimited quantities, but the exchange surfaces have to be large and become bulky when the heat losses to be removed reach some hundreds of kilowatts. Operation of the system then requires an expenditure of energy.

Water is only used when air cooling is not possible (e.g. in some indoor or underground installations of high power) or when water is readily available in large quantities at negligible cost. Water needs some power for its circulation, and precautions must be taken against freezing. When the losses are high, the use of water enables some low temperature (20–30°C) heat to be recovered, and the flow can be adjusted by reference to the outlet temperature of the water.

COOLING SYSTEM IN A TRANSFORMER

Most transformers are immersed in a liquid dielectric, and this type will be examined in detail, particulary the oil immersed transformer cooled by air. The dry-type transformer is simpler in principle if not in construction.

The transfer of heat from the active part to air takes place in two stages, firstly from the active part to the oil, inside the tank, and then from the oil to the air in the cooler. The movement of fluids takes place either naturally by convection or is forced by pumps or fans. Table 5.3 shows various types of cooling systems each of which have their own characteristics.

Natural circulation of oil and air

Temperature distribution is measured in the direction of heat flow and fluid flow. The section of a winding, Figure 5.2, shows the variation of temperature.

Table 5.3 Methods of cooling

Type of Cooling Agent	Symbol
Mineral Oil	O
Gas	G
Water	W
Air	A
Solid insulant	S

Type of Circulation	Symbol
Natural	N
Forced	F
Directed through the winding	D

Most Common Systems	Symbol
Dry transformer with natural cooling in air	AN
Dry transformer with forced air cooling	AF
Transformer with natural circulation of air and oil	ONAN
Transformer with natural oil circulation and forced air circulation	ONAF
Transformer with forced oil and air circulation	OFAF
Transformer with forced air circulation and directed oil flow through the windings	ODAF
Transformer with forced oil and water circulation	OFWF

At the interior of a winding the variation has a parabolic form, the heat being generated in a uniform manner and transmitted by conduction. At the surface of the windings, in the adjacent layer of oil, a sharp temperature drop occurs which is a function of the surface density of heat loss and the fluid velocity. Outside this zone, the temperature drops only little. A similar change occurs in the magnetic circuit and its cooling ducts.

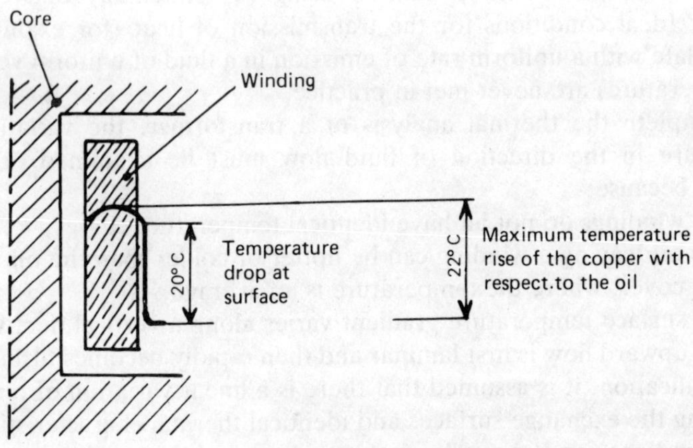

Figure 5.2. Variation of temperature in the windings

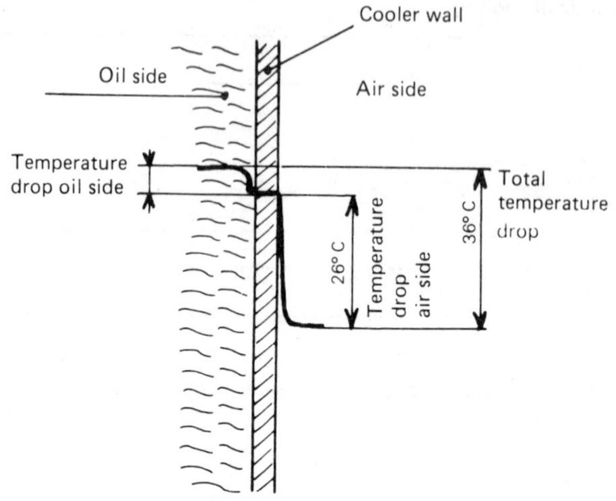

Figure 5.3. Temperature variation in a cooler

Figure 5.3 shows the temperature gradation through the cooler. The temperature drop is virtually concentrated in the oil layers at the surface of the metal wall. The heat is transmitted by conduction through the wall with negligible temperature gradient, and there is then a further temperature drop on the outside in the layer of air circulating along the wall.

The typical numerical values given in Figures 5.2 and 5.3 show that the surface temperature drop forms the major part of the variation. The efficiency of the cooler is represented by the total difference of temperature for given operating conditions and size.

The temperature drop is calculated by means of formulae derived from thermodynamics and fluid dynamics, using experimentally derived coefficients. Ideal conditions for the transmission of heat (for example, a vertical plate with a uniform rate of emission in a fluid of uniform velocity and temperature) are never met in practice.

To complete the thermal analysis of a transformer, the variation of temperature in the direction of fluid flow must be examined. This is complex, because:

(a) the windings do not all have identical temperature curves,
(b) the oil leaving a winding can be hotter or colder than the oil under the cover, where the temperature is an average,
(c) the surface temperature gradient varies along a vertical duct where the upward flow is first laminar and then rapidly becomes turbulent.

For simplification, it is assumed that there is a linear variation of temperature along the exchange surfaces and identical thermal characteristics for all the windings, and the manufacturer tries to achieve this to obtain optimum sizing.

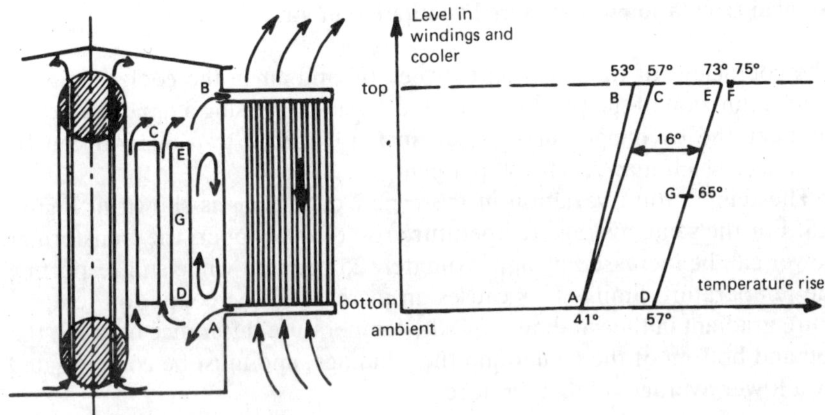

Figure 5.4. Natural circulation of oil and air. Temperature diagram: AB = oil in cooler; AC = oil in windings; DE = copper in windings; F = hot spot; AD = CE = copper-oil gradient; G = average temperature rise of copper

Figure 5.4 shows the variation of oil temperature in the windings and exchanger. The oil entering the exchanger is slightly cooler than that leaving the windings. The distribution of temperature in the winding parallels that of the oil in the ducts. The middle point corresponds to the value determined by variation of resistance, while the 'hot spot' situated in the top part governs the thermal behaviour of the insulation. The oil temperature at the inlet and outlet of the exchanger is also directly measurable. The temperature difference between copper and oil, and the average temperature of the oil in the winding ducts, are determined from the cooling-down curve of the windings obtained by measurement of the variation of resistance during a heating test.

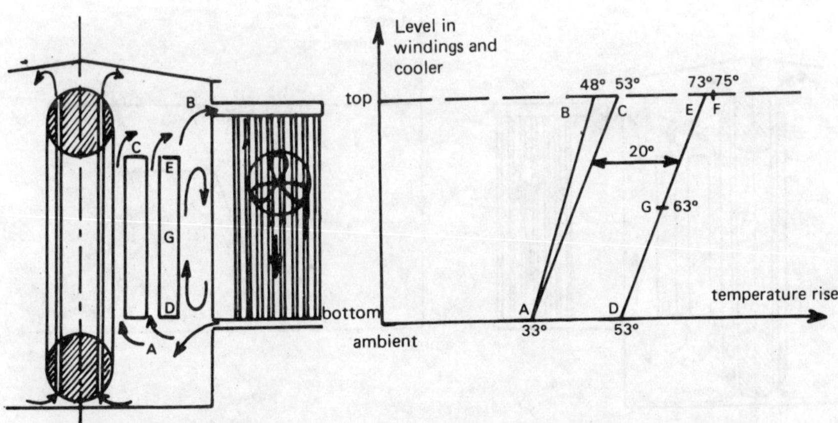

Figure 5.5. Natural circulation of oil, forced circulation of air. Temperature diagram (symbols as in Figure 5.4)

Natural circulation of oil; forced ventilation of air

The coefficient of transmission between oil and air in the cooler is essentially a function of the conditions on the air side. By using a fan to move the air near the exchange surfaces, an increase in the transmission can be obtained which may reach 100 per cent.

The temperature variation in these new conditions is shown in Figure 5.5. For the same average temperature rise of the copper, the transformer power can be increased by approximately 25 per cent while still respecting the temperature limits. This causes an increase in the copper/oil temperature gradient in the windings, and the temperature difference between the top and bottom of the cooler and the windings, and must be compensated by a lower average oil temperature.

Forced circulation of oil and air

More intense cooling of the oil increases its viscosity, restricting circulation of the oil and limiting the increase in power. To obtain better performance, an oil circulation pump must be used. The oil/air heat transmission coefficient in the cooler undergoes a further increase. The difference of temperature from top to bottom is reduced considerably (see Figure 5.6). The circulation of oil within the windings always occurs by convection. Most of the oil circulated by the pump flows in the tank around the outside of the windings.

The final step is to channel the oil leaving the cooler so that it flows towards the base of the magnetic circuit and the windings. The velocity of the oil in the windings is increased tenfold from cm/s to dm/s which

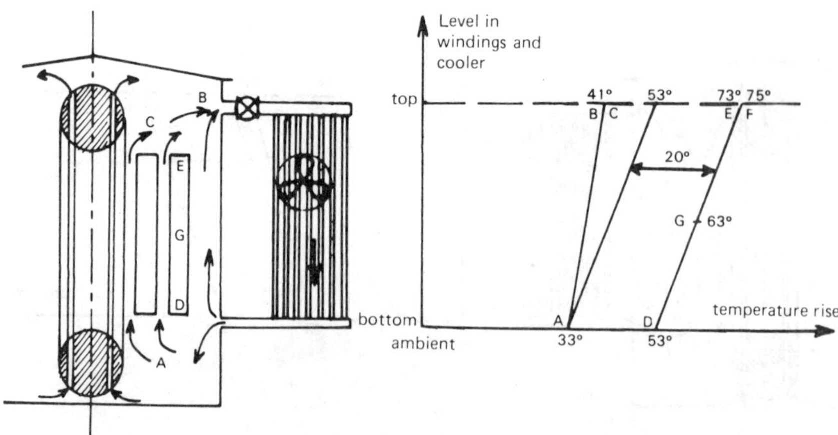

Figure 5.6. Forced circulation of oil and air. Temperature diagram (symbols as in Figure 5.4)

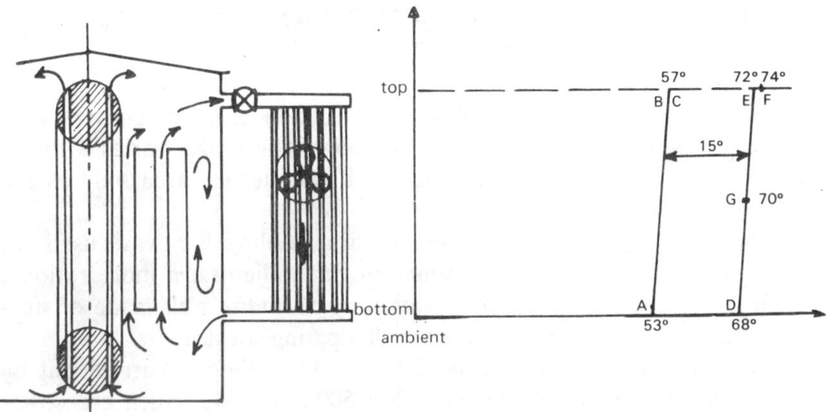

Figure 5.7. Directed oil circulation, forced circulation of air. Temperature diagram (symbols as in Figure 5.4)

practically doubles the coefficient of heat transmission. Figure 5.7 shows that the temperatures are very much more uniform, permitting an increase in the average rise of the copper by 5° (70°C instead of 65°C) without the temperature of the hot point exceeding the permissible limit.

This arrangement is used more and more in medium power transformers (from 10 to 100 MVA) and is almost universally used in high power transformers (above 100 MVA).

Figure 5.8. Naturally cooled distribution transformer (tank wall pleated to form radiators)

METHODS OF CONSTRUCTION AND OPERATING CHARACTERISTICS

As transformer power increases, their construction, from the point of view of cooling, becomes more and more complex. The power varies with the fourth power of the linear dimensions but the surface area only as the square of the linear dimensions.

At the bottom end of the range, natural cooling is always used. In addition to tanks provided with tubes or flat radiators, a more rational construction has been developed, consisting of a tank wall made of steel sheet folded into pleats to increase the dissipating surface.

Above a certain power, around 5000 kVA, cooling is carried out by radiators consisting of flat elements 300–500 mm wide which are either welded directly to the tank, or connected to it by means of flanged pipework, provided with valves, when the dimensions exceed the transport loading gauge (Figure 5.9).

For high power transformers up to 100 MVA, separate banks of radiators can be constructed, alongside the tank on an independent base. This

Figure 5.9. A 5 MVA naturally cooled transformer, with radiator banks attached

Figure 5.10. A 240 MVA transformer cooled by natural circulation of oil and forced circulation of air

technique is not often used as it requires a large ground area. As the power increases the radiators become bulkier, the connection to the tank causes problems and their output tends to decrease. The addition of fans enables the dissipation of losses to be doubled, and the size and weight of the transformer are reduced as a consequence (Figures 5.10 and 5.11).

A transformer designed to give its full power using forced ventilation can supply 50–60 per cent of this power using natural cooling when the fans are idle. This arrangement enables a saving to be made on the energy required to drive the fans during the periods of reduced power demand, or when the ambient temperature is sufficiently low. See Chapter 10, Ventilation of Cubicles. The fans can be started and stopped with the aid of thermostats which measure the oil temperature.

Fan noise can be a nuisance, but may be mitigated by correct attention to the profile and number of blades, the speed of rotation, and the suspension.

When size is critical, air blast coolers must be used. These are compact and standard assemblies consist of a fan and finned radiators. Mounted in sufficient number on a base fixed to the tank, they are fed in groups by circulating pumps with motors immersed in the oil, usually situated on the upper part of the transformer (Figures 5.12).

Where cooling by water is either a necessity (transformers for electric furnaces) or is an economic solution (hydroelectric power stations) the

Figure 5.11. Transformer with similar cooling to Figure 5.10 but with the addition of pumps

Figure 5.12. 20 MVA transformer cooled by forced circulation of oil and air (air cooling)

Figure 5.13. 40 MVA transformer cooled by forced circulation of oil and water (water cooling)

losses are removed by an oil/water exchanger of a standard tubular type. The water circulates through the tubes while baffles direct the oil across the outside surface of the tubes.

The circulation of both oil and water is forced, and the pressure of the oil in the cooler is always above that of the water so that if a leak occurs it is in the direction of oil into water. The metal used for the exchanger depends on the quality of the water used. Generally two or more exchangers are used with a total capacity above the value required for normal use to enable units to be changed over during operation. Pumps and exchangers fitted directly to the transformer increase the bulk of the latter very little (Figure 5.13).

COOLING CONTROL

Temperature measurement equipment becomes more complex as the power increases. All transformers are supplied with thermometers indicating the temperature in the top layers of oil, and the bulb is placed in a thermometer pocket in the cover.

With abnormally high temperature, the signal is given either by electrical contacts on the thermometer dial or by separate thermostats.

Transformers with forced ventilation operate generally with two or even three cooling stages, depending on the temperature. The cooling is natural while the temperature of the oil does not exceed 65°C. Above that a thermostat starts the fans, possibly in several stages, depending on the power attained.

When the oil circulation is forced, the pumps must be brought into use as soon as the transformer is put on voltage, and at the same time the first stage of the ventilation or the circulation of water in the water cooler must be started. For transformers of this type, no-load operation is impossible with the cooling stopped. As before, the equipment can be split into several sections, each being brought into service successively, depending on the oil temperature.

To enable the temperature of the windings to be controlled more directly, a device called a 'thermal image' has been developed, consisting of a resistor supplied by a current proportional to that in the windings and situated in a pocket in the oil in the top part of the transformer. This enables it to simulate the conditions at the hot spot. A thermal probe measures the temperature of the resistor which can be indicated by thermometer or thermostat. The thermal image is adjusted in the factory to suit the characteristics of the transformer.

CHOICE OF THE TYPE OF COOLING

This choice depends on the power, the local installation conditions and the method of operation.

Natural cooling is the most economic for small and medium power up to about 15 MVA. Artificial cooling equipment considerably increases the initial cost. Maintenance costs are the lowest with natural cooling.

Depending on the costs of the land and civil engineering works, it may be economically justified to use forced ventilation of radiators, between 10 and 100 MVA or even air blast coolers from 10 MVA, and in all cases above 100 MVA. The use of several stages enables the cooling to be adjusted to the power supplied and the ambient temperature. Cooling by water has to be specified in certain installations where the transformer is in a confined space (steelworks, underground power stations, etc). It is also a possible solution in other situations where water is already available in sufficient quantity and at a low price.

6 Overloads

INTRODUCTION

The characteristics of a transformer (losses, short-circuit voltage, voltage drop) depend on the rated values of the power, current and voltage at which the heating of the different parts must not exceed the limits set by the standard specifications.

These limits have been set after years of service experience to ensure that transformers may be operated permanently at their rated power and in the predicted ambient conditions for a normal service life of 20 years.

In practice such constant and consistent conditions are never experienced. The temperature of the various active elements of a transformer varies continually due to the daily and seasonal variations of the ambient temperature, due to the total losses, which vary with the power demand and determine the temperature of the liquid dielectric, and as a result of variations in the voltage and current which influence the temperature rise of the magnetic circuit and the windings above that of the liquid dielectric.

Thus in winter the temperature of the whole transformer, when out of service, can drop to −20°C or even less, while in summer the hot spot of the transformer windings under full load can reach 115°C.

The power demand fluctuates instantaneously, daily and seasonally. Its average value increases in the course of time due to the growth in consumption of electrical energy.

Thus the load on a transformer is sometimes lower, sometimes higher, than the rated value.

The overloads to which a transformer can be subjected are restricted by the effects of voltage drop on machines and other equipment, by the cost of

Figure 6.1. 20 MVA transformers installed at a private 90/5 kV substation

the losses, and by the ageing of insulating materials, resulting from a thermochemical phenomenon which is a function of the magnitude and duration of the overload.

Transformers immersed in oil will be examined in this chapter. The phenomena are the same for other types of transformers, and the transposition can be made from one to another, but the numerical values are specific to each type.

LAWS OF AGEING OF INSULATING MATERIALS

The insulating materials used in the manufacture of transformers, whether they are solid (paper, press board, wood) or liquid (oil), undergo a chemical alteration with time under the influence of heat and other agents such as oxygen and moisture.

The different materials present can influence each other. For example bare copper is a catalyst in the oxidation of oil while the products of the decomposition of paper accelerate the deterioration of oil.

Many important studies of this problem have been carried out in the last 40 years, involving long periods of testing. The effects have been felt in many fields; there have been developments in measuring techniques, changes in specifications, improvements in existing materials, creation of new products, and refinement of manufacturing processes.

The first clear usable formula related to the problem was developed by the American, Montsinger, author of the celebrated 'eight degrees rule'. Taking a 50 per cent reduction in the tensile strength of an insulating material as the criterion for the end of its life, he concluded that each 8°C rise in temperature reduced the life of the insulating material by a half. This is represented by the formula

$$t = t_0\, 2^{-\frac{\theta}{8}}$$

where t = duration of life

t_0 = constant (corresponding to the length of 'normal' life)

θ = the difference in °C from the normal temperature.

Montsinger's findings were developed by many other researchers. The introduction by Fabre of a chemical criterion, namely the degree of polymerization of cellulose (DP) for which a correlation with mechanical strength has been shown, has made it easier to study the phenomenon of ageing and the factors that influence it. The deterioration of cellulose appears as a rupture of the molecular chains which can be demonstrated by chemical measuring techniques.

Samples taken from old transformers show that cellulose insulating materials retain their properties well beyond the arbitrary limit set by Montsinger. Paper which has become brittle is still capable of resisting the normal electric stresses due to the impregnating oil and the barrier effect. Usually it is the oscillatory electrodynamic force during a short circuit which causes disintegration of the insulating material, followed by a breakdown between turns or coils.

The degree of polymerization, which is about 1300 for new paper, drops to about 150 for completely crumbly paper. Taking this as a more realistic value for the limit of the useful life of paper, Fabre obtained a constant of 5.5°C. During discussions in 1961 of the Transformer Working Group of CIGRE, 6°C was considered to be a more correct value for establishing a guide to loading of transformers.

Montsinger's law is in effect an approximation of a more general law, that of Arrhenius, which defines the speed of chemical reactions

$$t = A \cdot e^{-\frac{T}{B}}$$

where t is the time required for a given chemical reaction at the absolute temperature T, and A and B are constants.

The approximation is valid in the temperature range 80–130°C. Above 140°C new reactions appear in the decomposition of cellulose.

OVERLOADS APPLIED TO A TRANSFORMER

Operation of the transformer windings at a temperature less than 80°C, corresponding to a reduced load (60 per cent), or at a very low ambient temperature (0°C) causes negligible ageing of the insulating material. It is therefore possible, whenever necessary, and under well-defined conditions, to apply loads that exceed the rated power at which accelerated ageing of the insulating material is caused.

Depending on the system supplied, the transformer load diagram can vary widely. An example is that of factories operating partly or entirely with one, two or three substations.

Overloads are usually classed as:

recurrent overloads (normal load cycle) corresponding to planned or programmed power demands, repeating over a reasonable period of time, or

exceptional overloads (emergency conditions) due to unforeseen conditions such as transfer from a damaged transformer or accidental load peak.

Exceptional overloads are rare compared with recurrent overloads, and so greater and longer overloads may be tolerated, as the cumulative loss of life is not prohibitive.

It would seem that a measurement of the oil temperature of a transformer would be a simple and convenient method of checking its thermal state. However, this would be dangerous and unsuitable because the time constants for the heating of the windings and the oil are quite different, being several minutes for the former and several hours for the latter. The temperature of the windings rises much faster than that of the oil. Moreover, the heating of the windings relative to the oil is not constant and increases at low temperatures due to the increased viscosity of the oil. A thermal image consisting of a resistor immersed in the hot oil and carrying a current proportional to that in the windings is the most reliable, but it gives only instantaneous values.

The most practical and reliable method of determining a transformer overload capacity, in magnitude and time, as a function of the previous load conditions and the ambient temperature, is to calculate in advance the overload conditions for a large number of practical cases, based on standard transformer designs, and present them in the form of tables or graphs. Access to a computer is necessary for a calculation of this type.

THE IEC GUIDE TO LOADING

A guide to loading for transformers has been included for some time in the standards of some countries and in the operating instructions of some authorities generating power.

To bring unity to this issue, the IEC Study Committee No. 14 published a guide in 1972 applicable to transformers complying with IEC's publication 76 (which closely follows French standard UTE C 52–100). This guide has been approved by most large industrial nations. Its object is to help users operate their equipment and to make it easier to choose the nominal power rating for a given load regime.

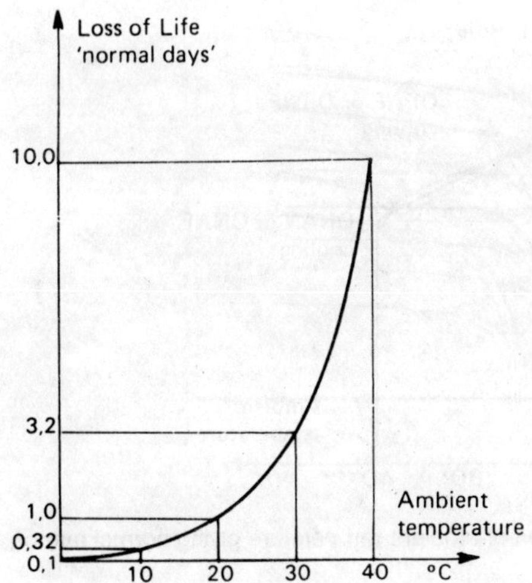

Figure 6.2. Loss of life after one day's operation on full load as a function of ambient temperature

Several preliminary maximum limits are fixed in accordance with the characteristics of the materials and the equipment, as follows:

150 per cent of the rated current for the current under normal cyclic conditions,

140°C for the temperature of the hot spot under all circumstances,

115°C for the temperature of the oil under exceptional conditions.

The characteristics of equipment such as terminals and tap changers can possibly justify even lower limits.

It is accepted that continuous operation of the hot spot at 98°C leads to a normal rate of degradation. This rate doubles for each increase in temperature of 6°C. The loss of life during overload or underload is expressed in 'normal days' corresponding to 24 hours operation at the condition given above (for the hot spot for a normal ambient temperature of 20°C).

From Figure 6.2, a 'normal day's' loss of life, for a transformer operating at full load, requires 10 days at an ambient temperature of 0°C but only 7 h 30 min at an ambient temperature of 30°C. Figure 6.3 shows how the power varies under continuous service conditions at various ambient temperatures. As can be seen, the old rule of 1 per cent of power for each degree was quite close to reality. A value of 0.8 per cent/°C would have been more exact.

The 6°C rule is shown in Figure 6.4, giving the number of operating hours corresponding to the same degradation as a 'normal day' as a function of the temperature of the hot spot. At 110°C, for example, the

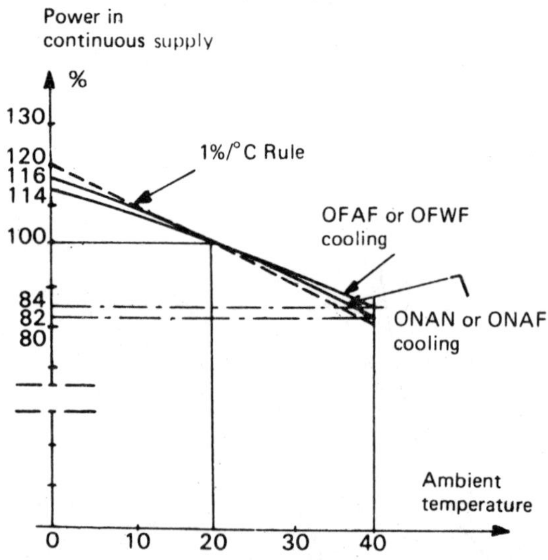

Figure 6.3. Continuous loading as a function of ambient temperature giving normal rate of loss of life

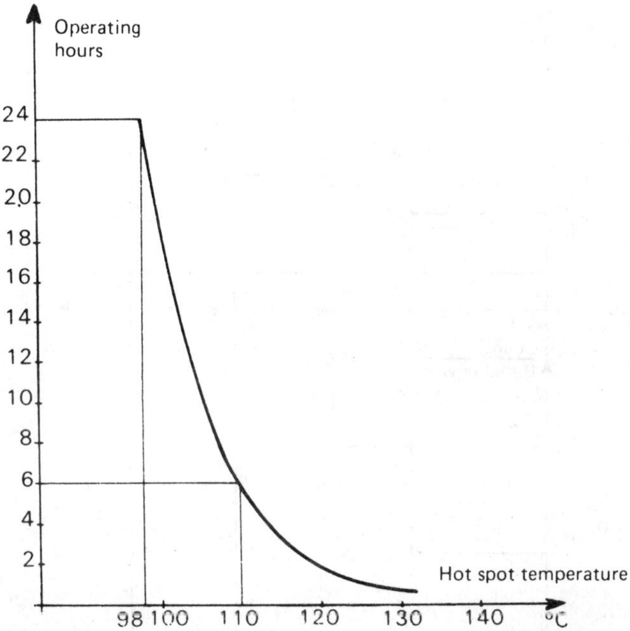

Figure 6.4. Operating time against hot spot temperature for one day loss of life

insulation ages four times as fast, the equivalent time being 6 hours. It is only 11 minutes at 140°C. Overloads in this range must be treated with caution.

Normal load cycles

The IEC Load Guide contains two groups of tables translated into the form of graphs, valid either for transformers with natural oil circulation and natural or forced air cooling (ONAN or ONAF) or for transformers with forced oil circulation and forced air or water cooling (OFAF or OFWF). These tables are based on a simplified load diagram, as a proportion of the rated power of the transformer and a duration of 24 hours. The initial load equals K_1 (see Figure 6.5), rises to K_2 for a period of t hours then drops to K_1 again. The ambient temperature can be any value between 0 and 40°C.

The value of K_2 is calculated so as to obtain an ageing equal to a 'normal day', the duration of the overload varying from 0.5 h to 24 h and the value of K_1 from 0.25 to 1.

Figure 6.6 shows the daily load cycles corresponding to ambient temperatures of 0–40°C for three overload durations: 0.5, 2 and 8 h and for the two types of transformers: ONAN, ONAF and OFAF, OFWF. The

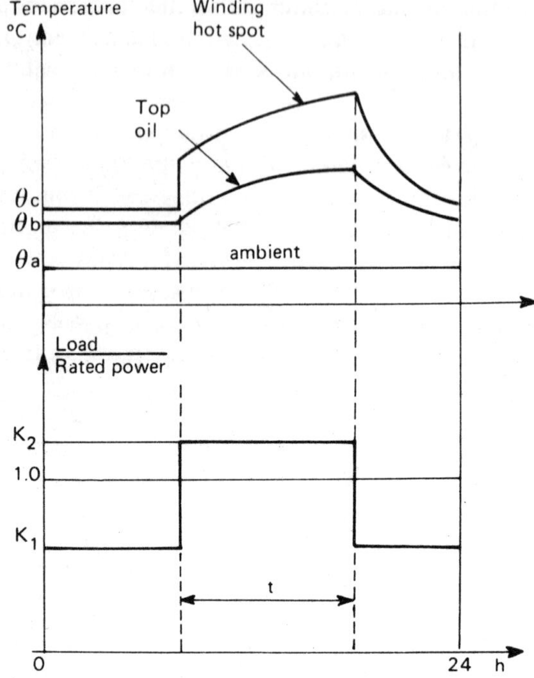

Figure 6.5. Typical load cycle and variations of transformer temperatures

ambient temperature and the duration of the overload have a very marked effect, whereas the base load has a decreasing importance as the overload duration increases.

The ambient temperature varies during a 24 h cycle. The value to be taken into account is that at the moment of the overload, when its effect on the ageing is much greater than during base load operation.

Take, for example, the overloads currently indicated for naturally cooled transformers constructed during the period 1945–1950. They are listed after a 25 year interval in Table 6.1, for an ambient temperature of

Table 6.1

Base loads (per cent)	Overloads (per cent)	
	1950	*1975*
50	130 (0.5 h)	150 (0.5 h)
	115 (2 h)	142 (2 h)
75	125 (0.5 h)	150 (0.5 h)
	115 (2 h)	130 (2 h)
100	120 (0.5 h)	—
	110 (2 h)	—

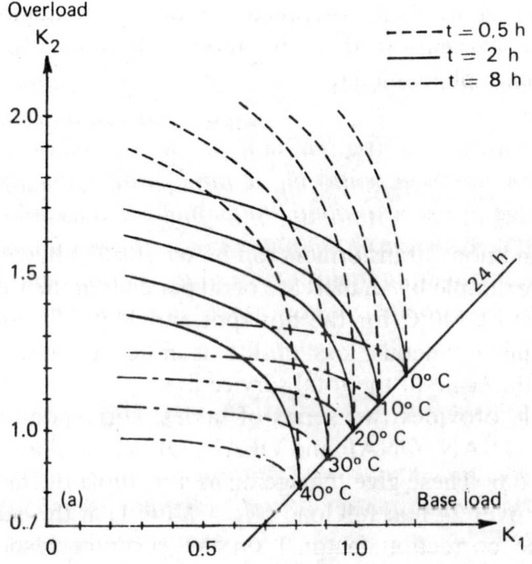

(a)

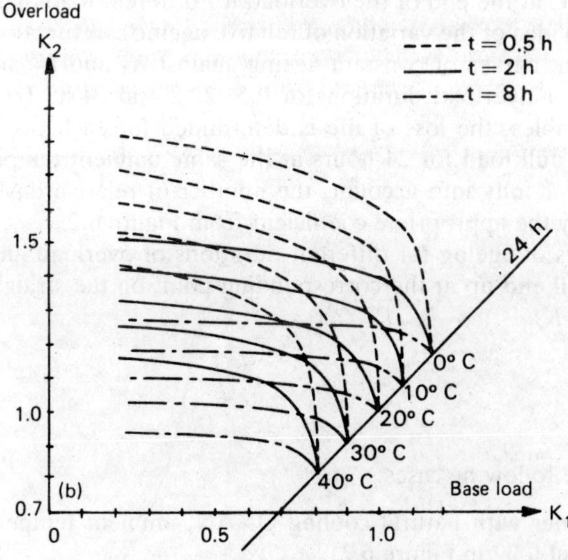

(b)

Figure 6.6. (a) ONAN, ONAF transformers (b) OFAF, OFWF transformers. Daily load cycles with overloads of 0.5, 2, 8 hours as a function of ambient temperature giving a normal day's loss of life

30°C. The overloads of 0.5 h are restricted to 150 per cent for normal service, but can reach 192 per cent under emergency conditions in the first case and 171 per cent in the second case. According to the IEC Guide, an overload of 120 per cent, 0.5 h is only allowed for a base load not exceeding 90 per cent. The differences in the table demonstrate the progress made in 25 years in the estimation of thermal effects and the knowledge of ageing phenomena.

Standby service

In exceptional conditions, transformers can be overloaded beyond the limits given above, for example by an increase of 50 per cent in the rated current, without the limits of 140°C for the hot spot and 115°C for the oil being exceeded. The instantaneous loss of life is much greater, but this is tolerable within the scale of the total service life.

The IEC Guide provides two series of tables, corresponding to transformers of types ONAN, ONAF and OFAF, OFWF, and a load diagram similar to Figure 6.6. These give the ageing as a multiple or fraction of that due to operation over 24 h at full load ($K_1 = K_2 = 1$) at the same ambient temperature. The correction factor 1 on the curve corresponds to the normal day's loss of life.

The code letters indicate the ambient temperature at which the hot spot reaches 140°C at the end of the overload for different load diagrams.

To give an idea of the variation of relative ageing as a function of a given load diagram, curves of constant ageing against K_1 and K_2 are shown in Figure 6.7 for overload durations of 0.5, 2, 8 and 24 h. To reduce the number of tables, the loss of life is determined for each case relative to operation at full load for 24 hours at the same ambient temperature. To take the latter fully into account, the number of relative days should be multiplied by the appropriate coefficient from Figure 6.2.

The curves of ageing for different durations of overload and the same loss of life all end up at the corresponding point on the straight line 24 h where $K_1 = K_2$.

Example

Consider the following cases:

1. Transformer with natural cooling ONAN, ambient temperature 10°C (coefficient 0.32 in Figure 6.2).
   base load: 100 per cent 23.5 h,
   overload: 160 per cent 0.5 h.

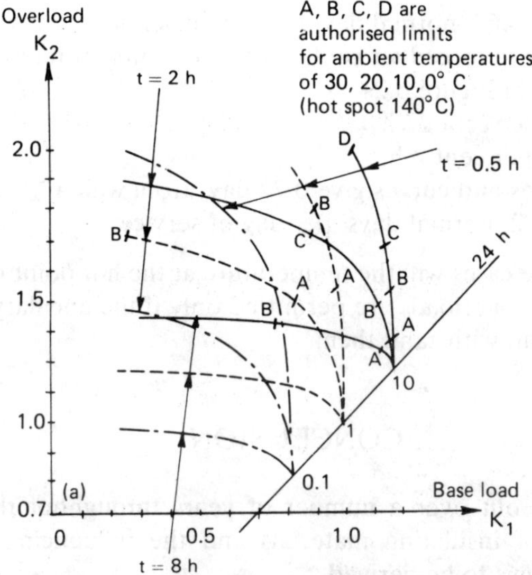

(a)

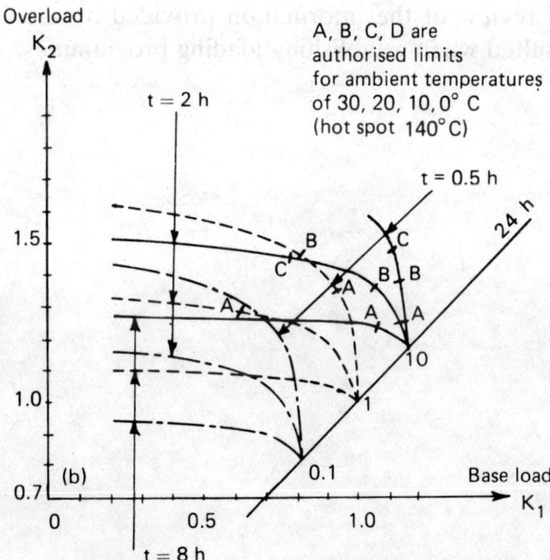

(b)

Figure 6.7. (a) ONAN, ONAF transformers (b) OFAF, OFWF transformers. Curves for constant loss of life relative to continuous operation on full load in the same ambient conditions

Table XI IEC Load Guide or interpolation on curve 7 gives 1.92 days, from which

1.92 × 0.32 = 0.62 'normal day' per day of service.
2. Transformer with forced air cooling ONAF, ambient temperature 40°C (coefficient 10 in Figure 6.2).
base load: 50 per cent 22 h,
overload: 140 per cent 2 h.
The same tables and curves give 0.22 day, from which
0.22 × 10 = 2.2 'normal days' per day of service.

In neither of these cases will the temperature at the hot point reach 140°C. These exceptional overloads are permitted only if the ancillary equipment and accessories can withstand them.

CONCLUSION

Research carried out over a number of years throughout the world on thermal ageing of insulating materials and the influencing factors has enabled certain laws to be derived.

The collation of a vast amount of theoretical and experimental knowledge by the IEC resulted in the Guide, which forms one of the most effective tools for the rational operation of transformers. Here, we give only a general review of the information provided by the Guide, which should be consulted when establishing loading programmes.

7 Operating limits of transformers and auto-transformers

BASIC PRINCIPLES

In a transformer the primary and secondary windings with voltages U_1 and U_2 have currents I_1 and I_2 flowing in opposite directions. In an auto-transformer the connections make it possible to use the primary winding to raise the

Figure 7.1. Active part of 250 MVA auto-transformer – 20/132/400 kv

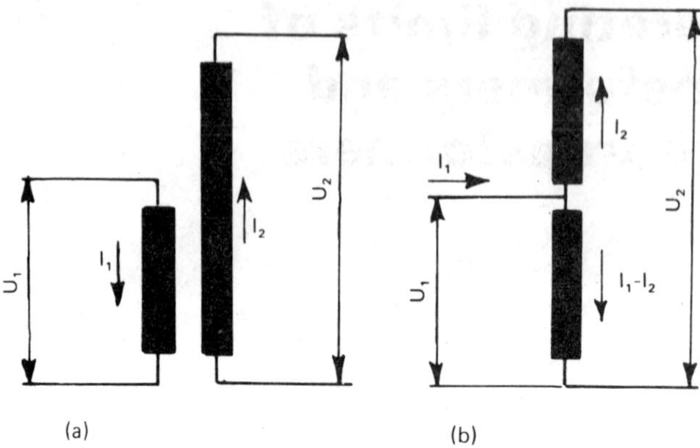

Figure 7.2. Operating principles of double-wound and auto-transformers. (a) Double-wound transformer (b) Auto-transformer

voltage in the secondary winding to U_1, enabling the secondary winding to be reduced $(U_2 - U_1)$. The current flowing in the primary winding is thus the difference of the two currents I_1 and I_2. There is as a result an obvious gain in terms of the active material and the size of the transformer (Figure 7.2).

Auto-transformers are used at low voltage, for distribution at 127 and 220 V, up to very high voltages for transmission at 420 and 765 kV. They have advantages and disadvantages which restrict their use compared with double winding transformers. We will be considering only three-phase units here.

Several types of auto-transformer exist depending on their use:

for interconnection between two systems of different voltages, possibly with regulation,
for regulation of transformer voltage within wide limits where the secondary is at low voltage, such as transformers supplying electric furnaces or rectifiers for electrolysis or traction,
for supply to synchronous or asynchronous motors at reduced voltage while starting.

Numerous connection arrangements are used for auto-transformers depending on the conditions of regulation: voltage, current, amplitude. Those most frequently used are shown in single phase in Figure 7.3.

Tappings at the neutral point, as in (a) of Figure 7.3, have the advantage of placing the tap changer in a zone less exposed to overvoltages, but they lead to a bigger transformer and are only used for small tapping ranges at very high voltage.

When the lower voltage varies and the higher voltage is fixed, the connections (b), (c) and (d) can be used, the choice depending on the

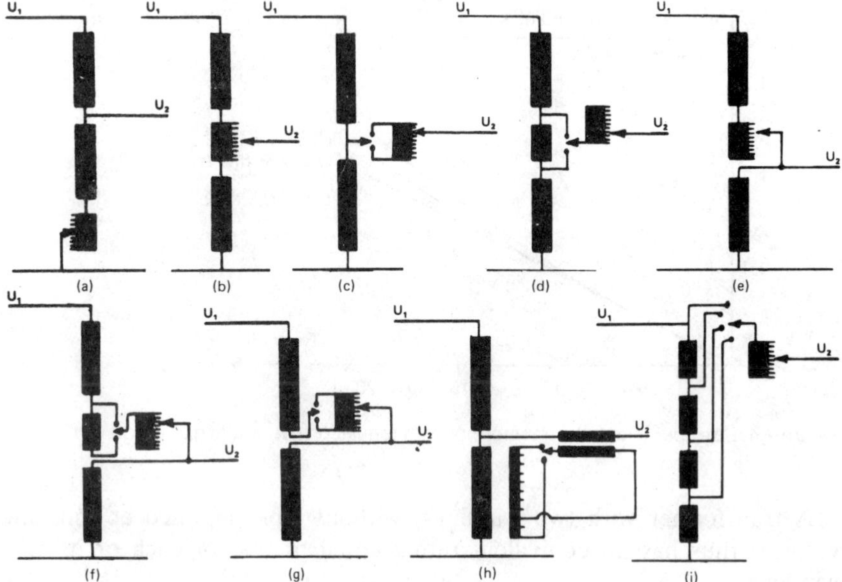

Figure 7.3. Principle methods of connecting auto-transformers. (a) U_1 or U_2 variable; (b), (c), (d), (h), (i): U_1 fixed U_2 variable; (e), (f), (g): U_1 variable, U_2 fixed

tapping range. When the higher voltage varies and the lower is fixed, connections (e), (f) and (g) are used and the tap changer is always placed on the low voltage side.

In special conditions where the performance of the existing tap changers is unsufficient (high current), the connection (h) must be used, comprising an auto-transformer with a tap winding and a separate transformer giving an additive or subtractive voltage on the low voltage side. The whole entity is clearly bigger, and this arrangement is used only in exceptional cases.

Regulating transformers for furnaces or rectifiers are generally of type (i) and comprise several steps of coarse regulation possibly right down to the neutral point.

Auto-transformers for starting belong to group (b).

EQUIVALENT SIZE

To establish a basis of comparison between transformers of different characteristics such as power, regulation and windings, the notion of the two-winding equivalent is used. For a winding or part of a winding, the power rating is the product of the maximum current and the maximum voltage in service. For the complete transformer the two-winding equivalent is half the sum of the power ratings of all the windings.

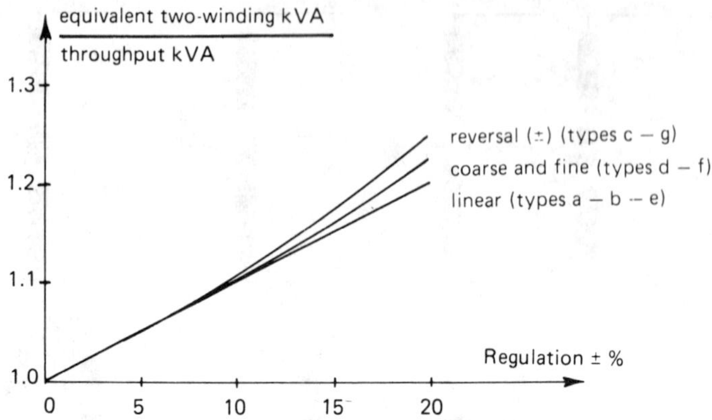

Figure 7.4. Size coefficient of a transformer with regulation (types a to g are shown in Figure 7.3)

A transformer with two windings, without taps, supplied at constant voltage, thus has an equivalent rating equal to that of each of its two windings.

As soon as regulation is introduced on one of the windings, and full output is required at all the tappings, the equivalent two-winding rating increases by virtue of an additional tap winding.

Figure 7.4 shows the variation as a function of the percentage tapping range relative to the postion for the two types of connection most used, namely reversal of the tap winding, and fine and coarse windings.

This concept does not take into account influential factors such as the service and test voltages, the number of windings, or the connections and types of windings, but it does provide an approximate estimate of the quantity of active material required for a given transformer, and it is a very convenient method of evaluation.

An auto-transformer with a fixed transformation ratio U_1/U_2 ($U_1 < U_2$) is the simplest case. Its size relative to a double-wound transformer of the same output is

$$p = 1 - \frac{U_1}{U_2}.$$

The reduction in size increases as the voltages U_1 and U_2 approach one another. An auto-transformer of 2000 kVA, 15/20 kV, without tappings, has a size of

$$2000(1 - \tfrac{15}{20}) = 500 \text{ kVA.}$$

The introduction of voltage regulation, requiring a variable number of additional windings, as shown in Figure 7.3, does affect the calculation

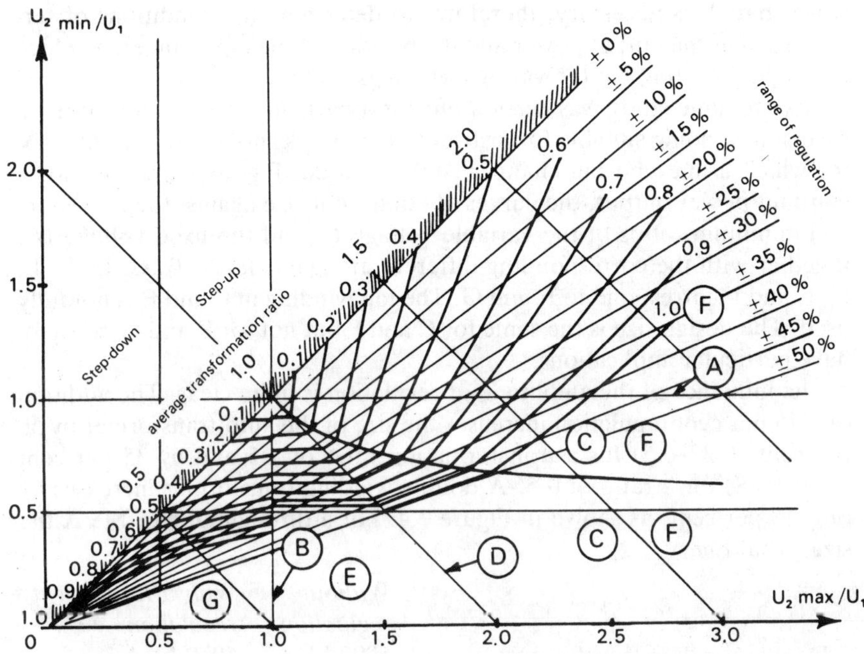

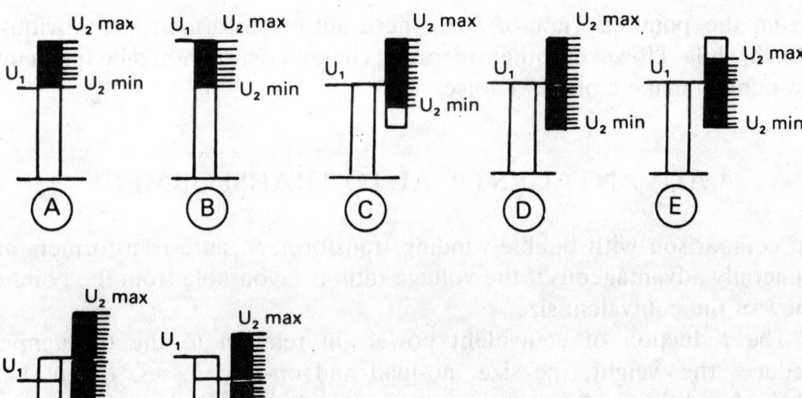

Figure 7.5. Curves relating maximum and minimum voltage ratios with constant value of $\dfrac{\text{equivalent two winding kVA}}{\text{throughput kVA}}$ as parameters (reversing tap winding)

somewhat. It is necessary, therefore, to determine the conditions of service, such as the output power and the position of the tap changer, at which the highest currents in the various windings occur.

A systematic study was carried out for a particular connection, using a reversing winding suitable for distribution networks not exceeding 72.5 kV for which a line end tap changer could be used. Figure 7.5 shows, for a constant power output, the curves of equivalent size against the maximum and minimum ratios of the variable voltage U_2 and the fixed voltage U_1, together with the corresponding different arangements A, B, C, D, E, F, G. C and F correspond to E and G. The tap winding in C and E is not fully used. The design size is the same for C and F but not for E and G, as these have particular applications.

The influence of the amplitude of regulation is quite clear. The addition of ± 15 per cent regulation increases the size of the auto-transformer by 50 per cent (0.33–0.5) for a transformation ratio of 1.5 and by 35 per cent (0.50–0.68) for a ratio of 0.5. A double-winding transformer increases by only 18 per cent, as shown in Figure 7.4. For a throughput of 5 MVA the sizes would be:

	Without regulation	*With ± 15 per cent regulation*
Double-winding transformer:	5000 kVA	5900 kVA
Auto-transformer 20/30 kV:	1667 kVA	2500 kVA
Auto-transformer 20/10 kV:	2500 kVA	3380 kVA

From the point of view of size, these auto-transformers are obviously worth while. However, other operating characteristics should be taken into account to make a proper choice.

ADVANTAGES OF AUTO-TRANSFORMERS

In comparison with double-winding transformers, auto-transformers are generally advantageous if the voltage ratio is favourable from the point of view of the equivalent size.

The reduction of equivalent power, in relation to the throughput, reduces the weight, the size, no-load and on-load losses, the no-load current and the short-circuit impedance.

If P is the power of a transformer, the linear dimensions vary as $P^{0.25}$ and the weight and volume as $P^{0.75}$, other things being equal. Thus a gain of 50 per cent in the equivalent rating reduces the weight, the losses and the no-load current by 40 per cent.

The short-circuit impedance also goes down with the equivalent rating, which is an advantage from the point of view of voltage drop. However, when regulation is incorporated, it varies considerably, with the tap position and also with the relative position of the windings.

For connections between two systems at very high voltage, where the power transmitted is in the order of GVA, the use of auto-transformers makes it possible for high power equipment to be constructed as a single, transportable unit. This would not be possible with normal transformers.

DISADVANTAGES OF AUTO-TRANSFORMERS

Generally star connections are used. Delta connections, which cause a variable phase difference between the two voltages, and star/zigzag are used infrequently. The two systems must have the same neutral point conditions; earthed directly or through an impedance, or isolated.

An electrical connection between the primary and secondary systems is not always desirable, particularly when the voltages are quite different. If the neutral point is not solidly earthed, the lower voltage side can be subjected to a high potential if there is a fault to earth on the high voltage side. In practice, auto-transformers are only used on distribution systems where the neutral is connected to earth.

The auto-transformer is particularly sensitive to atmospheric over-voltages. The series winding is much shorter than the high voltage winding of a double-wound transformer, but both are subject to the same over-voltages. The tap winding, with its tap changer placed preferably near the medium voltage terminal, is subjected to much harder treatment than its counterpart which is nearly always placed near the neutral point of the double-wound transformer. The auto-transformer therefore requires a much higher standard of insulation than a normal transformer. Protection by surge arresters is especially required.

As the reactance of an auto-transformer is relatively low, short-circuit currents can reach critical values. It is essential to take account of the incoming and outgoing systems to obtain realistic values of the reactance. A particularly important case is that of a single-phase short circuit to earth. If there is a tertiary winding connected in delta, it is subjected to consider-able currents. To limit these, special measures may have to be taken to increase the zero-sequence impedance (reactance in the neutral, or in the delta). The design of the tertiary in this case raises mechanical consider-ations rather than thermal ones.

PROBLEMS OF THE TERTIARY WINDING

It has been a longstanding practice to provide star/star connected trans-formers and auto-transformers with a tertiary winding in delta with the following aims.

1. Stabilizing the phase-to-phase voltages in the case of an unbalanced load (e.g. single-phase load between one phase and neutral). Without a

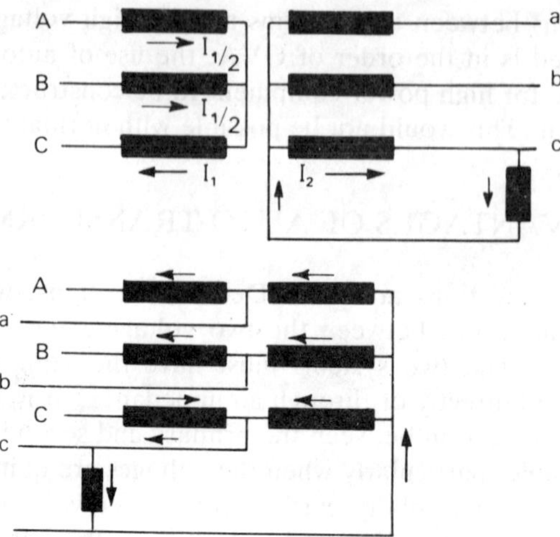

Figure 7.6. Current distribution due to single-phase loading in a star/star transformer and in an auto-transformer

tertiary winding (Figure 7.6) the current flowing in the uncompensated phases is purely magnetizing, and by saturation causes deformation of the phase voltages, displacement of the neutral point and heating of the tank due to the stray flux losses. The addition of a winding in delta whose rating is a third of that of the transformer (Figure 7.7) balances the ampere turns in all three phases, eliminating these various phenomena.

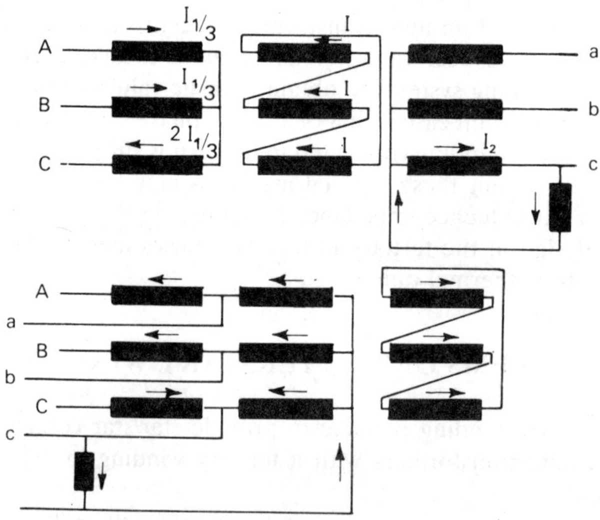

Figure 7.7. Current distribution due to single-phase loading in a star/star transformer and in an auto-transformer with delta connected tertiary

2. Suppressing the third harmonics due to the no-load current in the earth connections when the neutral point is earthed. These harmonics induce disturbances in neighbouring low voltage cables, particularly telephone lines that are not screened. In the case of an isolated neutral, harmonics develop in the voltage and the flux causing oscillation of the neutral point. A tertiary delta suppresses these faults. Moreover the general use of grain-oriented steel for the magnetic circuits keeps the no-load current to a minimum. The disruptive effect of the harmonics is no longer very noticeable.

3. Reducing the zero-sequence reactance so as to obtain a ratio of less than three for the zero-sequence reactance/positive-sequence reactance, which is characteristic of a system connected to earth. A tertiary winding in delta reduces the zero-sequence reactance close to the short-circuit reactance, by 5–15 per cent, whereas it can reach 100 per cent for a three-phase core with three legs and even higher values for single-phase cores, three-phase cores with five legs, and shell types.

4. Supplying reactances to provide the reactive energy of the system, or to supply station auxiliaries or a local system.

In the first three cases, the design of the tertiary is more often governed by the necessity to withstand the short-circuit forces which are due to the reduced reactance between this winding and the neighbouring main winding. Generally this is not a problem for the transformer. On the other hand, the tertiary of an auto-transformer with earthed neutral is likely to be subjected to considerable currents during a phase-to-earth fault in one of the connected systems. It is advisable to increase either the reactances or the area of the tertiary conductors.

The addition of a tertiary winding of power $\frac{1}{3}P$, where P is the through-put power, increases the cost of a transformer by about 10 per cent. For an auto-transformer, depending on the voltages, this increase can be up to 50 per cent. A tertiary winding, therefore, should be considered only if it is absolutely necessary.

A study should be carried out for each case, taking into consideration the various points raised above. Usually a transformer whose power does not exceed some tens of MVA, with a three-legged core, has no need of a tertiary winding in delta. Auto-transformers have operated satisfactorily like this for a number of years in North America and Europe.

SPECIAL AUTO-TRANSFORMERS

Certain industrial processes require voltage regulation over a wide range, often between zero and maximum voltage. As a general rule the power decreases with the voltage in accordance with a rule unique for each application. These applications are electric arc or resistance furnaces, rectifiers for electrolysis, and rectifiers for supplying traction motors.

The voltage variation is obtained by an auto-transformer connected as in Figure 7.3(i) supplying the main transformer. The size depends on the required variation of power with voltage. For example, for voltage variation at constant secondary current, the equivalent two-winding power is a little less than 50 per cent of the throughput power.

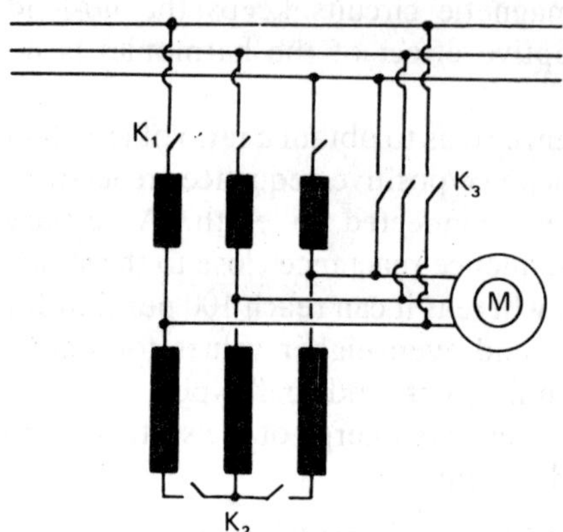

Figure 7.8. Auto-transformer for three step starting of a synchronous motor

Another particular application is the supply at reduced voltage to large synchronous or asynchronous motors during starting to avoid too high a power demand. Figure 7.8 shows the arrangement of connections for three-step starting without interruption of power supply.

In step 1, the switches K_1 and K_2 are closed, K_3 open, and the motor starts under reduced voltage.

In step 2, the switch K_1 is closed, K_2 and K_3 open. The neutral point is opened, the series winding remains in circuit and restricts the current.

In step 3, the switch K_3 is closed, K_2 open, K_1 open or closed, and the motor is at full voltage.

Auto-transformers such as these which operate for a few minutes only, have flux densities in the iron and current densities in the copper much higher than normal transformers. The surface of the tank is generally sufficient to dissipate the heat generated without exceeding the temperature limits prescribed by the standards. Off-circuit tappings are provided to adjust the starting voltages.

CONCLUSION

In comparison with normal transformers of the same characteristics, the auto-transformer is more advantageous in terms of physical size, but it is subjected to more demanding conditions of service.

Apart from the special applications referred to above, where its use cannot be questioned, the auto-transformer should be chosen only after a detailed examination has been made to ensure that it conforms to all the conditions of service.

As a general rule such a solution can be chosen if the following characteristics are grouped together:

system connected to earth,
system of restricted short-circuit power,
sheltered situation with respect to over-voltages,
transformation ratio not far from unity (0.5–2),
balanced load.

8 Protection

THE OBJECT OF PROTECTION

The transformer is one of the most reliable elements in an electrical system. However, although the stresses under permanent operating conditions can be easily assessed for sizing the transformer, the stresses in transient conditions (electrical, electrodynamic or thermal) may vary widely and must be analyzed statistically. The possibility of operating faults must be taken into account, as well as their consequences, and the means by which they can be dealt with.

An interruption to supply due to a fault can have negligible consequences or it can be catastrophic depending on the system supplied and the duration of the fault.

The protection of a transformer has three aims:

1. To protect it from external disturbances, short circuits, over-voltages and overloads.
2. To protect the systems connected to it and the environment from the effects of the fault in the transformer itself.
3. To monitor the operation of the transformer, warn against incipient faults and restrict damage if a fault does occur.

Protection has to be considered from an economic point of view, taking into account the probability of a particular type of fault, its possible consequences (loss of production, cost of repair, material damage, etc.) and the cost of the protection required to restrict it.

It is obvious, therefore, that the policy must be different in each case. The most advanced protection systems are used on transformers providing

Figure 8.1. 25 MVA transformers installed at a 150/15.75–21 kV substation (forced cooling with radiators provided with fans)

an essential supply in exposed conditions. More simplified devices can be used for installations consisting of several sources of supply and serving non-priority sections of the system.

Some devices can fulfil several functions. However, one function may require several pieces of equipment.

EVENTS OUTSIDE THE TRANSFORMER

These are likely to be over-voltages resulting from atmospheric phenomena (lightning) transmitted by overhead lines. The presence of a length of cable is, in this case, likely to create problems through wave reflection. Switching in the system can produce over-voltages of less steep but longer duration, stressing both liquid and solid dielectrics. These over-voltages must be restricted in amplitude to a value below the transformer withstand level.

Short circuits subjecting the transformer to currents of 10–20 times the rated current have thermal and electrodynamic effects. A transformer is normally designed to withstand some tens of dead short circuits, lasting no more than 2 seconds, during its life. If there are likely to be more, then a special construction is required. The short circuit must be eliminated more rapidly as the current intensity increases.

Overloads arise from planned or fortuitous circumstances. In the first case, temperature increase of the insulating material must not exceed standard values. In the second, a certain time limit can be tolerated but it will mean some loss of life of the transformer.

INTERNAL FAULTS IN A TRANSFORMER

These can be very diverse in nature:
(a) dielectric: shorting between windings or between live parts and earth, partial discharges;
(b) electric: bad contacts in leads or bad contacts in the tap changer;
(c) electrodynamic: forces due to external or internal short circuit;
(d) electromagnetic: eddy currents induced in the magnetic circuit or the clamping structure, the tank;
(e) thermal: abnormal temperature rise, hot spot, thermal ageing or pollution of oil;
(f) mechanical: vibrations, leakages or defective operation of the tap changer.

The different types of defects have degrees of gravity corresponding to the amount of damage they cause and the consequences. Some of them develop slowly (e.g. vibrations, partial discharges) and do not immediately endanger the equipment. Having had warning, it is possible to defer the remedial work until a suitable opportunity arises.

Other defects require safety measures to be taken (e.g. overloads), allowing a certain time for them to be implemented. Defects likely to cause substantial damage (e.g. short circuits, internal breakdown) require immediate and automatic action.

METHODS OF PROTECTION

The basic requirement of any protection method is to be able to discern between abnormal conditions and normal service functions, such as switching and allowable variation of current, voltage and frequency, to which there should be no hurried reaction.

There are various operating principles and methods of application. Some act directly on the disturbance and reduce it to a safe level; these are, for example, spark gaps and lightning arresters for external over-voltages, fuses for over-currents, diaphragms or non-return valves for excess internal pressures, and fire protection against an oil fire.

Others measure a characteristic magnitude of the fault: current, temperature, quantity of gas and, depending on the seriousness of the fault and the operating requirements, isolate the transformer or bring in an alarm. Methods used in this case are differential protection, earth leakage and Buchholz relay to signal internal faults, over-current relays to safeguard against overloads and short-circuits, thermometers and temperature probes to show temperature rise of the oil and windings, and monitoring equipment for the auxiliaries such as tap changer, pumps, fans, etc.

These different systems will be reviewed, dividing them according to the type of fault into three categories: external over-voltage, over-currents and overloads, internal faults.

Protection against external over-voltage

For each rated voltage there is a corresponding withstand level against external over-voltage (IEC 76–3).

Thus for a rated voltage of 63 kV (maximum service voltage 72.5 kV) the impulse withstand voltage level is 350 kV, while for a rated voltage of 400 kV (maximum service voltage 420 kV) there is a switching over-voltage level of 1050 kV associated with the lightning impulse level of 1425 kV or 1300 kV, and a switching over-voltage level of 950 kV associated with a lightning impulse of 1175 kV.

The installation must be equipped with means to reduce the over-voltages occurring in service to values less than the specified ones, with a sufficient margin to spare.

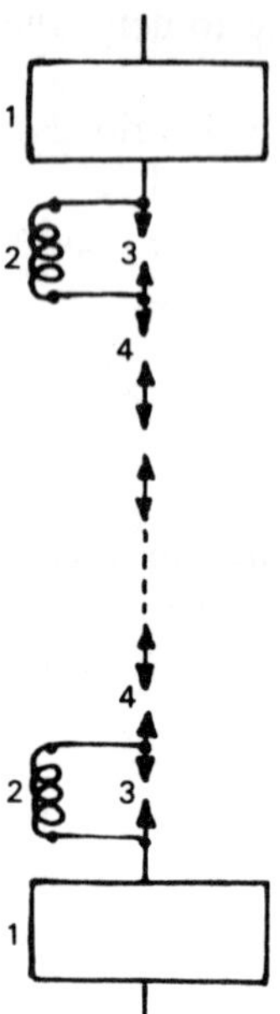

Figure 8.2. Arrangement of a surge limiter. 1. Non-linear resistance. 2. Coil. 3. Coil shunt gap. 4. Quench gap

The best protection is afforded by lightning arresters. They have to pass currents of some thousands of amperes for a hundred or so microseconds due to the atmospheric over-voltages, and currents of several hundred amperes for a period of a millisecond due to switching over-voltages. They are selected for the level of protection they must afford, and the rated voltage, which is the maximum continuous voltage at which they can operate continuously and at which they will clear the follow-through power frequency current.

Two types of lightning arrester are currently in use. The oldest consist of non-linear silicon carbide resistors in series with spark gaps which spark over under the effect of an over-voltage then break the resulting power frequency current. The more recent consist of non-linear zinc oxide resistors without a spark gap. For a given service voltage, the latter provide protection at a voltage level lower than that provided by arresters with spark gaps.

Protection against over-currents and overloads

For overload protection, the standard thermometer mounted on all transformers can be fitted with contacts to sound an alarm and to switch for an abnormal temperature or a dangerous oil temperature. This signal needs to be supplemented by information from other instruments, as it gives no information on the temperature of the hot spot in the windings.

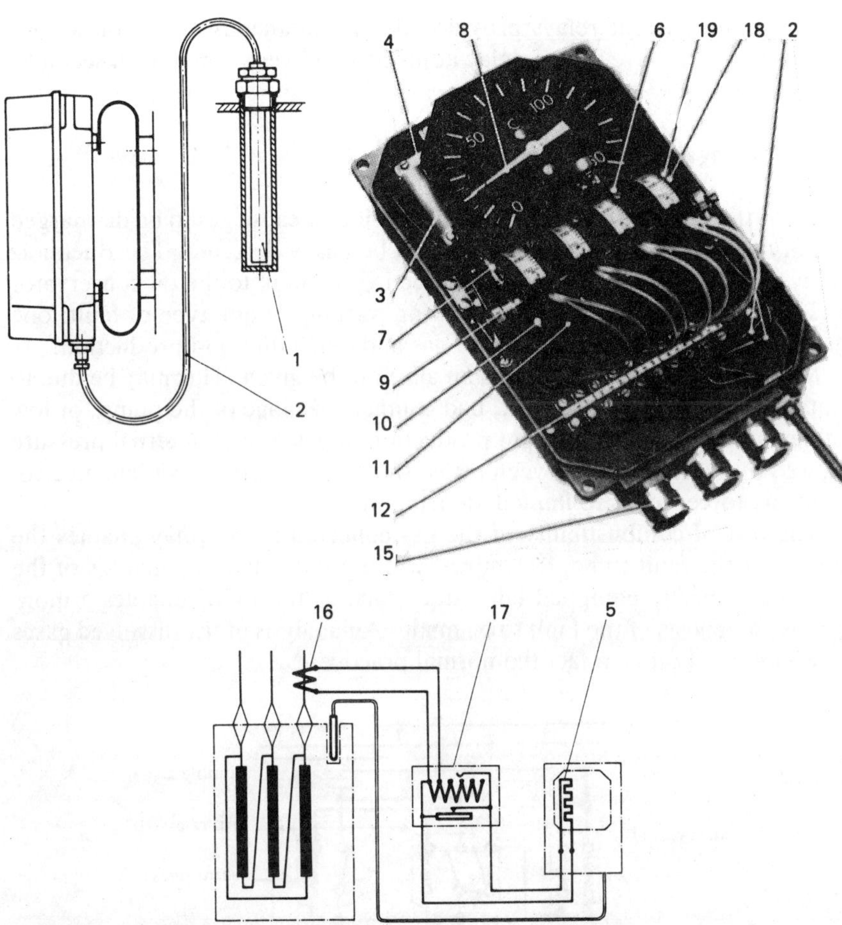

Figure 8.3. Thermal image. 1. Oil-filled pocket. 2. Connecting pipework. 3. Measuring bellows. 4. Compensating bellows. 5. Heating resistor. 6. Heating resistor wires. 7. Lever mechanism. 8. Pointer. 9. Spindle. 10. Mercury switch. 11. Calibrated drum. 12. Terminal block. 15. Cable gland. 16. Line-current transformer. 17. Adaptor unit. 18. Needle. 19. Clamping screw.

The thermal image (Figure 8.3) is the best method of checking the loading condition of the transformer. This relay, immersed in oil at the upper part of the transformer tank, comprises a resistor through which flows a current proportional to that in the winding being monitored. The temperature rise and the time constant are adjustable. The thermal image can thus monitor the temperature of the hottest point in the winding and can emit a signal when the allowable limit has been exceeded. Indirectly, depending on the transformer load, it can switch in or out the steps of the cooling equipment as required.

Maximum current relays provided for instantaneous operation at currents of 3–6 I_n, have a time delay adjustable between 0 and 6–10 seconds.

Protection against internal faults

Most of the internal faults mentioned previously cause gas to be discharged at varying rates and quantities; this can be easily detected. The Buchholz relay mounted on the pipework connecting the tank to the oil conservator has two floats with contacts each giving warning of one type of fault: one associated with slow production of gas and one with rapid production.

Slow production of gas causes an alarm to be given. This may be due to hot spot, ionization, core fault, bad contacts, leakage of the pump, or low oil level due to leakage. Rapid production of gas or high internal pressure caused by short circuit between turns, shorting to earth, or violent electro-dynamic forces leads to immediate tripping out.

The test of combustibility of the gas collected in the relay enables the nature of the fault to be determined without difficulty. An analysis of the gas in a specially equipped laboratory such as the LCIE enables a more precise diagnosis of the fault to be made. An analysis of the dissolved gases in a sample of oil is in fact the normal practice.

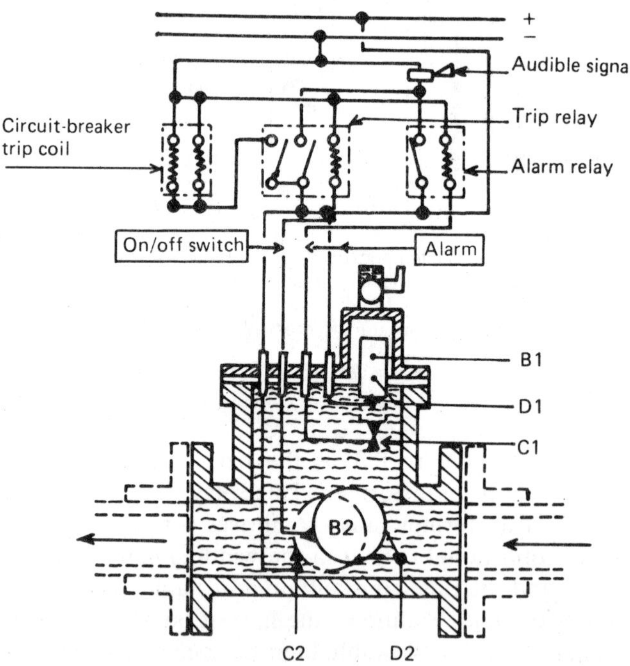

Figure 8.4. Schematic of a Buchholz relay. B1, B2. Metal floats. D1, D2. Pivots. C1, C2. Mercury switches

The robust design and the simple and reliable operation of the Buchholz relay have led it to be widely used (Figure 8.4).

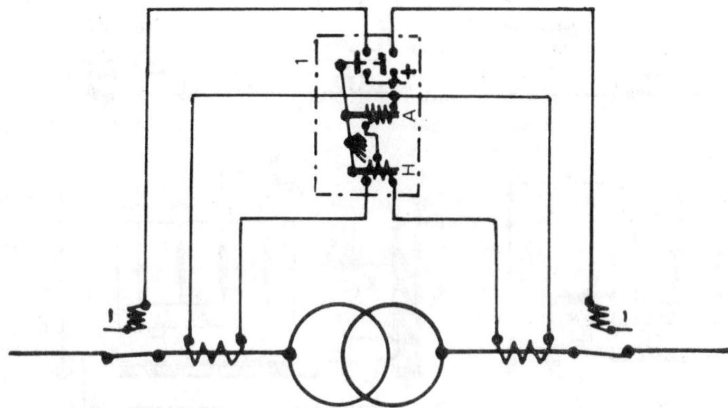

Figure 8.5. Principle of differential protection of transformers. 1. Differential relay. A. Trip coil. H. Compensating coil

Differential protection, which is purely electrical in nature, compares the currents entering and leaving a transformer. It detects the short circuits between windings and turns, and shorting to earth, of an effectively earthed system.

Although it is applied successfully to alternators, differential protection causes problems with transformers:

It is necessary to have current transformers with characteristics as similar as possible, in spite of the different currents and insulation levels.
Compensation is required for possible different phase angles on the primary and secondary sides.
The sensitivity is reduced of a transformer with on-load tap changing.
There may be disturbances due to switching-in currents.

In fact these difficulties can be overcome, and differential protection is a good method of detecting internal short circuits. However, due to its high cost and the fact that it does not show up faults in the magnetic circuit and the points of abnormally high temperatures, the Buchholz relay used in conjunction with an earth leakage relay is preferred in many countries.

In systems where the neutral is not isolated, a fault to earth in a transformer shows as the circulation of heavy currents to earth from the tank wall. As the transformer is insulated from earth (tank, pipework, cable sheaths, etc.) a special current transformer connected between the tank and the earth supplies a current metering relay which signals the passage of a fault current, and gives the necessary command to trip (Figure 8.6).

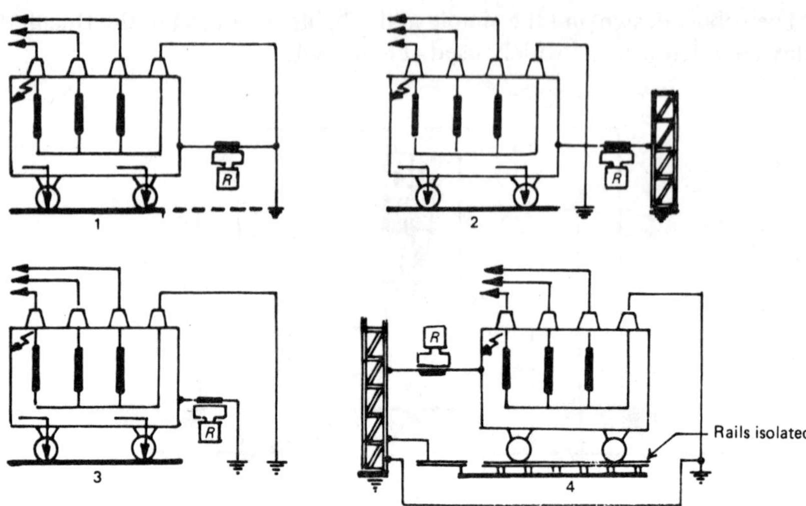

Figure 8.6. Tank protection (see Figure 8.5)

Other protective devices listed below are not required to detect faults but only to restrict their effects. Major faults release a large quantity of energy, vaporizing the oil and causing high pressure surge waves. To avoid or to limit deformation of the tank, it is current practice to provide transformers either with a pressure relief diaphragm or a non-return valve, fitted with pipework to allow the oil to discharge in a safe area.

If a bushing ruptures or the tank splits after a violent internal fault, ignited oil can escape from the transformer. Fire protection can be of three types:

a sprinkler system to spray water on the transformer,
a system blowing nitrogen under pressure through the oil in the lower part of the tank, after partial emptying,
a spray with a fine mist of water.

These devices are usually automatic and tend to cool the oil to bring it below the temperature of ignition and so extinguish the fire. The fine water spray is simple and efficient and is used more and more.

SPECIFICATIONS

French Standard Specification NFC 13–200 governs high-voltage installations and prescribes the protection that must be fitted to power transformers:

Internal faults:
device to detect the emission of gas (Buchholz relay),

pressure relief diaphragm or safety valve.
Internal breakdown to earth:
  earth-tank leakage protection.
Over-voltages:
  lightning arresters in all cases or spark gaps in less exposed positions,
  with low resistance earthing resistor.
Short circuits:
  circuit-breakers or fuses or current operated relays.
Overloads:
  two temperature detectors, alarm and trip, or current operated relay.
Auxiliaries:
  control of circulation of liquid dielectric and cooling fluid.

Most of the protective devices described in this chapter are thus prescribed by the Specifications. The choice of type and in certain cases the number depend on the installation conditions: connections, powers, voltages. A comparison should be made in each case between the probability of faults, their costs and their accepted risks, and of the cost of protection.

9 Installation conditions

DESIGN OF A TRANSFORMER SUBSTATION

The connection between an electrical energy supply system and a load or distribution system is made through a transformer substation, in which is situated all the equipment necessary for the operating network to meet instantaneous demands and for ensuring control and protection of the system and equipment.

Design of the substation must conform to the detailed regulations contained in the relevant codes and standards. The principal component, being the bulkiest and the heaviest, is the power transformer. Local service conditions have an effect on the siting of the station, the construction of the transformer and its installation conditions.

The design must take into consideration the safety of personnel, the protection of the equipment, trouble free operation of the distribution system, and continuity of supply. Transformer substations may be indoor, outdoor or have a prefabricated enclosure. The drawings of the installation must be submitted for the prior approval of the distributor.

The regulations deal essentially with civil engineering works, HV electrical installations and LV electrical installations. The standards for transformer substations apply to either consumers' indoor substations with a capacity less than or equal to 2.5 MVA with a voltage not exceeding 57 kV, or high-voltage electrical installations, generally outdoor, in which the voltage is less than 66 kV.

In addition, the Centre d'Equipement du Réseau de Transport de l'Electricité de France (equivalent of CEGB Transmission) has prepared a Specification and Technical Conditions Relative to the Construction of

Figure 9.1. Single-phase 360 MVA, (400/√3) kV generator transformers

Transformer Stations and their Annexes, constructed under its control or with its approval.

CIVIL ENGINEERING WORKS

For low power indoor substations, the building must be situated adjacent to a public or private road so that it is accessible at all times and to allow easy handling of all equipment, particularly the transformers. The floor must be designed for dead and live loads.

The high-voltage circuits must be made inaccessible by placing them sufficiently far away from walkways reserved for staff, or by using partitions or screens.

The transformer cell must include positions for the rails on which the transformer runs, ducts for HV and LV cables and openings for the earthing connections. When an oil pit is included, its volume must be related to the volume of oil in the transformer, be provided with a gravel infill to extinguish the oil, and an outlet pipe. To prevent pollution, a soakaway pit must be avoided and the pit must be leak proof.

To ventilate the station, the lower air inlets should be more than 20 cm above the floor and preferably behind or under the transformer. The air is evacuated by means of chimneys, skylights or openings discharging into open air. The design of the ventilation circuit is dealt with in Chapter 10.

In basement or underground substations, the ventilation and cable ducts must be arranged so that water cannot get in.

The exterior works associated with larger transformer stations comprise, for each transformer, a plinth with

ground beams supporting the runway rails,
drainage pit for removing the oil,
plinths for auxiliaries such as the water coolers and pump sets,
connection between the service track and the transformer emplacement,

Figure 9.2. 630 kVA dry type transformer in a pre-fabricated enclosure

Figure 9.3. 100 MVA, 220 kVA transformers in an outdoor substation

foundations for the protective walls between transformers,
if required, acoustic enclosures.

To ensure satisfactory operation of the earth leakage protection, there must be no contact between the transformer runway rails and the reinforcement of the support plinth, and isolating gaps must be provided between the transformer emplacement rails and the service rails.

If the emplacement is geographically separated from other buildings and accessible only to electricians, no special personnel protection measures are needed. In all other cases, the provision of an oil pit is not required when a Buchholz relay is fitted to the transformer. However to safeguard the equipment from the spread of fire following a fault, the provision of an oil pit and also sand in the ducts leading from the cell is strongly recommended.

ELECTRICAL INSTALLATIONS

For transformers with parts in open air, which are live under normal conditions (e.g. bushings and connections to busbars), or are likely to be subjected to high potential due to internal or external fault (e.g. tank and auxiliaries), their position in the station is subject to the regulations already referred to. These cover two principal aspects, installation distances and earth circuits.

Installation distances

For medium-voltage indoor substations (5.5–30 kV), the partitions between cells and the doors must have a minimum height of 2 m (2.3 m in certain cases) so that bare conductors and live parts are out of reach of people at ground level. The minimum distances in air are indicated in Table 9.1 (taken from French Standard UTE C 13–100).

For installations at higher voltage, indoor or outdoor, the distance from the floor to the lower part of the insulators must not be less than 2.25 m whatever the voltage. This may require low power transformers in outdoor substations to be raised or the bushings to be mounted on extensions on the tank cover.

The distance of the connections above ground level depends on the rated voltage. For 30 kV or less, the distance is 2.5 m; between 30 and 45 kV, it is 2.6 m; above 45 kV it is 2.8 m.

The minimum installation dimensions between conductors or between conductors and earth are given in Table 9.2 (extracted from Standard UTE C 13–200).

Table 9.1 Minimum distances in air

Rated voltage of distribution systems (kV):		5.5	20	30
Corresponding insulation level of the equipment (kV):		7.2	23	36
Description	*Distances in sketch below*	*Minimum distances (cm)*		
Outdoor buildings for stations connected to overhead systems:				
Centre line dimensions, station entries		24[1]	47	54
Centre line dimensions between anchorages,				
horizontally		50[1]	75	100
vertically		75[1]	100	150
Inside buildings:				
Height of insulating supports (smooth or fluted) between the part under tension and earth[2]		9[1]	16	30
Centre line dimensions, conductors and fuses[3]	a	24[1]	30	39
Distance between parts under tension connected to different phases[2,3]	b	10[1]	16	30
Distance between high-voltage conductors and low-voltage conductors not protected by a duct or a metallic screen connected to earth[4]		27[1]	45	65
Distance between parts under tension and the faces of cell walls or permanently erected screens or earth	d	9[1]	16	30
Distance between parts under tension and the faces of wire netting cells (or screens) permanently erected		30	30	30
Distance between parts under tension and access doors or screens not permanently erected		30	30	30

Note: These centre line dimensions have been chosen so that they correspond to those shown in the equipment standards. Except in special cases it is recommended to comply with them.

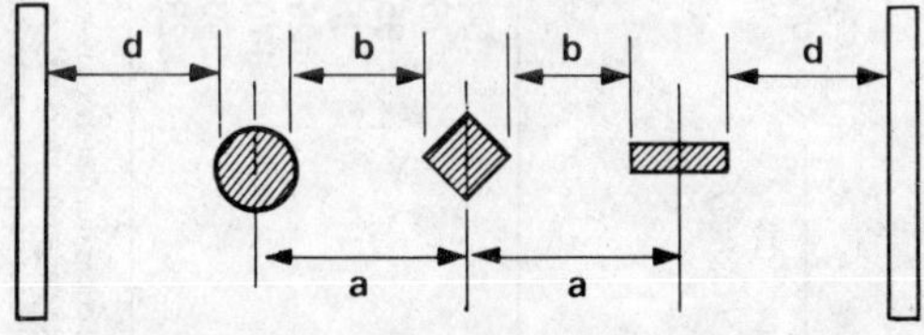

[1] This distance can be increased to the corresponding value in the 20 kV column by application of the note in clause 3.1.1.2.
[2] Distances conform with the requirements of Norme UTE C 10–100. They apply provided they comply with clause 3.1.3.2 of the present standard.
[3] Does not apply to terminal boxes, or to other equipment such as isolators or circuit-breakers for which the current regulations require impulse tests. This is also valid for transformers (impulse tests according to the standards).
[4] Does not apply to conductors at their connections to transformers.

Table 9.2 Minimum installation distances

Rated voltage		Impulse test voltage N_t (peak kV)	Distance between live parts and earth corresponding to voltage N_t (cm)	Minimum installation distances (cm)				
Installation U_n (kV effective)[1]	Insulation level U_i (kV)			between bare live parts and earth or active parts at low tension[1]		between bare live parts of different phases		Centre line distances, bare conductors
				indoor	outdoor	indoor[2]	outdoor	indoor
5.5 or 6.6	7.2	60	9	9	10	10	minimum 30	24
10	12	75	12	12	13	12		24
15	17.5	95	16	16	18	16		30
20	23	95	16	16	18	16		30
20[3]	24	125	22	22	22	22		30
30	36	170	32	32	33	32	38	39
48	52	250	48	50	53	50	61	—[4]
66	72.5	325	63	65	70	65	80	—[4]

[1] For wire mesh partitions and doors the distances between bare live parts and mesh or doors is equal to U cm with a minimum of 30 cm (U being the rated voltage expressed in kV); in the case of hand railing the distances from bare parts must exceed 2 m.

[2] These values are also imposed for centre line distances between fuses.

[3] The values in this line must be applied in installations where the neutral is isolated or is connected to earth through a high impedance (installation $1++$).

[4] The minimum value is the width of the live parts increased by the distance between phases.

Earth circuit

In each substation there is an earthing circuit to which must be connected

the earthed metal of all HV and LV circuits (cable sheaths and screens),
the metal tank of the transformers (possibly with a current transformer
for earth-leakage protection),
the surge arresters or HV spark gaps.

The neutral of the LV circuits is connected to the LV earth connection,
directly, by an impedance, or by an over-voltage limiter (in the case of non-
earthed LV circuits). This earth circuit can be common with the preceding
one when the earthing resistance is sufficiently low, but it must be separate
if the earthing resistance is not low (see Table 9.3).

Table 9.3 Values of earth resistance (Ω)

Earthing circuits joined:	$\leq$ 1
Earthing circuits separated:	
overhead or mixed system	$\leq$ 10
underground system	$\leq$ 3

The quality of the earth connection is of prime importance. Too high
resistance can cause dangerous increases in potential in the earthing conductors, in the case of a phase-earth fault, on the high voltage side. For LV
systems, the type of the protection system depends on the earthing resistance value, in accordance with Table 9.3.

Below the values shown, HV protective spark gaps can be connected to
the substation earth; above, separate earth connections or surge arresters
are required. An earthing resistance exceeding 30 Ω renders surge arresters necessary.

INSTALLATION OF THE TRANSFORMER

Before the arrival of the transformers on site, the necessary documents,
general arrangement drawings and erection instructions will have been
collected and studied in detail.

The exact position must be checked as well as the positions of accessories
and cable panels. The equipment required for offloading, setting in position and erecting must be available at the required time as well as all the
equipment for treating and checking the oil (if necessary). An electricity
supply must also be available.

For industrial transformers, road transport by lorry is used for the
smaller units, low load trailers for the medium sized units and special
trailers for the large units. Rail transport uses similar equipment. The

Figure 9.4. Three-phase 150 MVA, 210/66 kV transformer on a low load trailer

criteria for the choice are the minimum gauge for the route envisaged, the weight, the obstacles en route and the availability of access for unloading.

The transformer may be despatched completely erected and full of oil, therefore practically in working order, if circumstances allow. The accessories exceeding the transport loading gauge, such as the bushes, coolers, and control cabinet, may be dismantled and sent separately, the tank being partly emptied of oil.

On arrival and before being unloaded, a close examination is made of the equipment to reveal any possible damage resulting from the move and to avoid any future dispute with the transporter. Certain delicate accessories, such as the bushings, must be unloaded with the greatest care.

Figure 9.5. A single-phase 360 MVA transformer being erected

The transformer is usually transferred directly to its site where the erection and commissioning are carried out. If this is not possible, the accessories must be stored in the dry, and the parts for the oil circuit carefully sealed. The air dryer or the air or dry nitrogen system mounted on the tank must be checked regularly.

Transformers despatched empty of oil should be filled as soon as possible with treated oil, this operation being carried out under vacuum for voltages equal to or greater than 60 kV. The oil should reach a level about 150 mm below the cover in such a way as to allow the subsequent erection operations, but making sure that the windings and their insulation are covered.

The erection instructions, which should be closely followed throughout, will give details of the method of erecting the bushings, the cooling system, the auxiliaries, accessories, and control cabinets and panels.

The bushings, particularly large ones, must be handled very carefully. After removal of the blanking plates covering the openings during transport, the flexible conductor or tubular conductor used for the bushings is bolted to the cover, and the transformer connections are made.

The on-load tap-changing equipment and its control gear are subject to special instructions. Their erection is preferably carried out by a specialist.

All the elements of the cooling system: radiators, pipework, oil conservator must be given a thorough cleaning with oil to ensure that they are perfectly clean and dry.

During the filling of the cooling circuits, further oil will be added if necessary to the transformer tank to avoid any active part being exposed to air should the oil level fall.

Similar care must be taken with erection of auxiliaries, pumps, valves, Buchholz relays and air dryer. One of the valves at the bottom of the tank is used to fill up to the mark on the level indicator. For voltages equal to or greater than 60 kV the operation is carried out under a vacuum of about a millimetre of mercury. The oil must be thoroughly filtered, degassed and dried prior to its use in the transformer.

On completion of the erection the transformer is ready to be put on voltage. After connection of the HV and LV bushings to the station busbars and completion of the wiring of the auxiliaries to the control panel, the transformer can be put into service.

COMMISSIONING

A complete set of checks and tests must first be carried out, as follows:

Tank and cooling system
    opening of all valves on the cooling system,
    direction of rotation of pumps and fans,
    operation of flow indicators for the circulation of oil and water (if necessary),

gas release on bushings and Buchholz relay and check on the operation
of the latter,
closure of the bypass on the automatic non-return valve of the conservator oil reservoir,
oil level,
oil and air leaks,
check on the oil, dielectric strength (oil testing cell),
operation of the fire protection system.

Electrical circuits
agreement between the connections and the signals on the indicator
panel,
measurement of the transformation ratio on all taps,
measurement of the insulation resistance of the windings,
correct mechanical functioning of the tap changers,
adjustment of the spark gaps for the protection of the bushings,
check on the earth circuits,
continuity of wiring circuits.

When these tests have been completed the transformer can be switched in.
In the case of water or air cooling, the transformer cannot operate permanently on no-load, with the auxiliaries stopped. The oil and water circulation pumps and the fans are automatically started on closure of the
circuit-breaker.

When tension is first applied, it is preferable to increase the voltage
progressively. The transformer is left on volts, with no load, for several
hours before its load is applied in steps. When several stages of cooling are
involved, there will be particular instructions to be consulted.

As a transformer may take one or two hours to heat the oil, several
hours are required under constant load and constant cooling conditions for
the thermometer to indicate the oil temperature under continuous load and
at the ambient conditions pertaining.

10 Ventilation of enclosures

COOLING CONDITIONS FOR TRANSFORMERS INSIDE BUILDINGS

The power of transformers is defined in accordance with the IEC recommendations for the maximum temperatures of the dielectric and the windings under standard ambient conditions. These are:

minimum temperature: above $-25°C$,
maximum temperature: below $+40°C$,
average daily temperature: below $+30°C$,
average annual temperature: below $+20°C$.

These conditions apply to transformers installed indoors and outdoors. Only transformers installed in enclosures which are difficult to ventilate, or those installed underground, are considered as special cases, and must be subject to particular agreements.

In theory, heat is transmitted from a naturally cooled transformer to air by radiation and convection. However the shape of the coolers, consisting of closely stacked flat plates, is chosen to obtain the minimum volume of cooler for a given power, so the transmission of heat by radiation is very small. For a transformer installed in an enclosure, the radiated energy heats the walls of the enclosure, which are often constructed of a thermally insulating material, and a major part of this energy is returned to the ventilating air inside. It may be considered therefore that all the transformer losses are removed by convection.

Transformers installed outside are subjected to conflicting influences. Some, such as rain and wind, assist cooling; others, such as solar radiation,

Figure 10.1. 1000 kVA transformer installed in an enclosure

do not. Temperature variations for transformers installed inside are more regular, and depend on the enclosure size and its ventilation.

Tests carried out in Britain have shown that unless the air inlet and outlet openings are correctly placed relative to the enclosure shape and transformer position, the flow of air can give rise to eddies which seriously reduce the efficiency of ventilation.

For distribution transformers, natural ventilation of the enclosure based on the chimney effect is generally adequate. The difference in weight of the column of hot air inside the enclosure and the column of air outside creates the motive force to overcome the head losses at the required rate of flow. The necessary height of the enclosure above the transformer increases with the losses. Above a certain power it becomes economically and technically preferable to use forced ventilation with a fan operating either continuously or near to full load.

The air inlet at the lower part of the enclosure is situated either in the wall near to ground level and in the vicinity of the transformer, or,

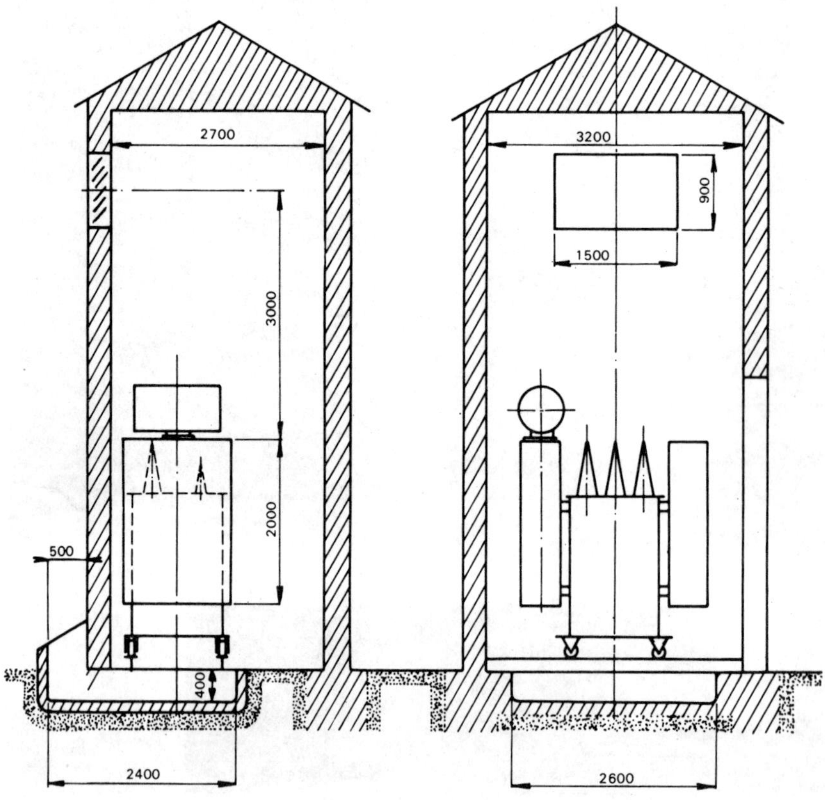

Figure 10.2. Cross-sections of a transformer enclosure

preferably, immediately under the transformer using an underground duct to introduce the air from the outside.

The outlet openings are situated at the top of the enclosure and are provided with shutters. A chimney may be necessary to create a column of hot air high enough for convection (Figure 10.2).

The position of these openings is important. The air inlets, particularly if they are metal, should not be exposed to the full sun. If they have to be placed on the south side, a screen should be provided to keep them in the shade. It is preferable to avoid placing the ventilation openings on the side facing the prevailing wind. The use of an exhaust system of the blower type, which operates in any wind direction, is the best solution.

If, as is recommended, the transformer is accessible on all its sides, the area of the air passage in line with the radiators poses no problems.

ENCLOSURE DIMENSIONS

Instructions for the installation, commissioning and maintenance of power transformers contain a set of curves (Figure 10.3) enabling an approximate determination of the enclosure size to be made. The parameters are:

maximum temperature of outside air: 35°,
temperature rise of the air due to heating: 15°,
flow of air: 0.1 m^3/s/kW.

Knowing two of the three terms: transformer losses, area of ventilation openings, and updraught height, it is possible to determine the third. The updraught height is the vertical distance between the centre of the radiator and the centre of the top shutters. For example, a transformer having losses of 20 kW must have an updraught height of 5 m for ventilation openings of 1.5 m^2.

This method gives a quite good approximation and makes it possible to check that an installation is adequately sized. Where the dimensions are close to the limits, it is preferable to carry out a more accurate calculation using analytical methods.

NATURAL VENTILATION OF ENCLOSURES

There are three stages in the calculation: the flow of air necessary, the updraught pressure, and the area of the air passage.

Flow of air

Air normally contains moisture in variable quantities. As a result its density is lower and its specific heat higher than dry air. If we take dry air

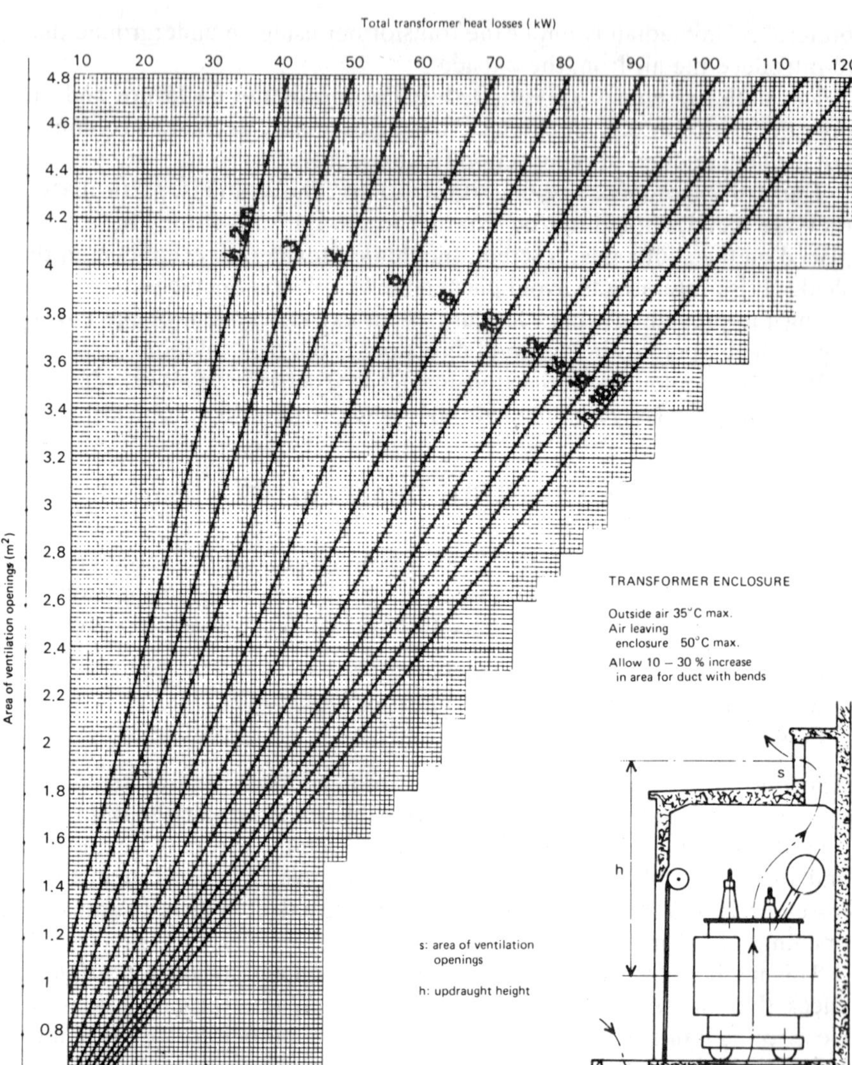

Figure 10.3. Ventilation duct areas

as the basis for the calculations, the higher energy requirement to cause movement and the higher flow provide a margin of safety.

The characteristics of air used for the calculations are:

$$\text{density} = 1.293 \times \frac{p}{760} \times \frac{273}{273 + t_E} \ (\text{kg/m}^3),$$

specific heat $c = 0.24$ (kcal/kg°C),

p = atmospheric pressure in mm Hg,

t_E = temperature of incident air °C.

The quantity of air at temperature t_E and atmospheric pressure H, required to remove 1 kW of losses for a temperature rise $(t_S - t_E)$ is thus

$$Q = \frac{1}{3600} \times \frac{860}{0.24} \times \frac{1}{1.293} \times \frac{273 + t_E}{273(t_S - t_E)} \times \frac{760}{H} \ (\text{m}^3\text{/s})$$

from which

$$Q = 2.82 \times 10^{-3} \times \frac{273 + t_E}{t_S - t_E} \times \frac{760}{H} \ (\text{m}^3\text{/s})$$

Table 10.1 gives the flow of air in m³/s per kW of loss for a pressure of 760 mm of mercury and several values of the inlet air temperature and the temperature rise of the air. It would appear that the temperature of the air entering has only secondary importance (increase of flow of 0.4 per cent per °C) relative to the permissible rise in air temperature.

A transformer of 2000 kVA 20/5.5 kV having total losses at full load of 24 kW requires an air flow of 1.37 m³/s, at 30°C and 760 mm mercury, for a temperature rise of 15°C.

Table 10.1 Air flow in m³/s per kW loss

Air temperature at inlet t_E (°C)	Air temperature rise $(t_S - t_E)$ (°C)			
	10	15	20	25
−20	0.0713	0.0476	0.0357	0.0285
−10	0.0742	0.0494	0.0371	0.0297
0	0.0770	0.0513	0.0385	0.0308
10	0.0798	0.0532	0.0399	0.0319
20	0.0826	0.0551	0.0413	0.0331
30	0.0854	0.0570	0.0427	0.0342
40	0.0883	0.0588	0.0441	0.0353

The temperature rise of the air must be such that the temperature rise of the oil and the winding does not exceed the values prescribed in the relevant standards. Figure 10.4 shows the variation of the temperature rise of the cooling air, the oil and the windings with the temperature of the incident air for a naturally air cooled transformer in the open.

From this diagram it would seem that a temperature rise of 25–30° is also acceptable for a transformer installed in an enclosure. The conditions for renewing the air around the transformer are, however, less favourable as shown in Figure 10.5.

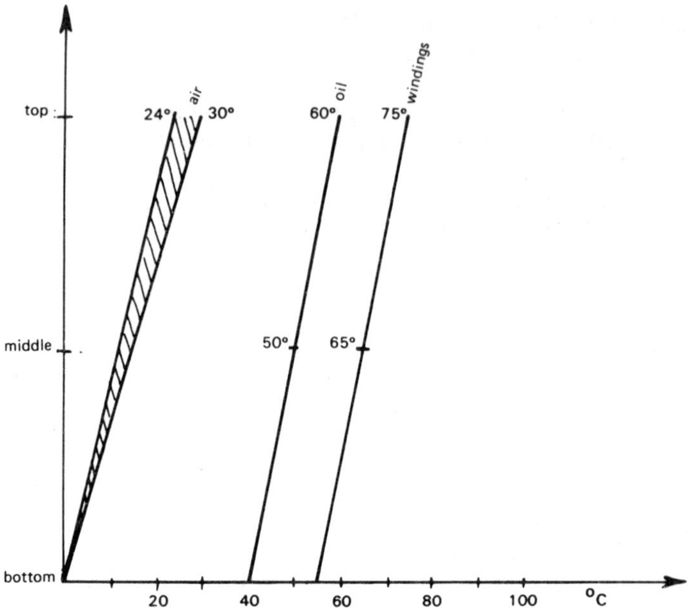

Figure 10.4. Variation of temperature rise above ambient between the top and bottom of a naturally cooled transformer

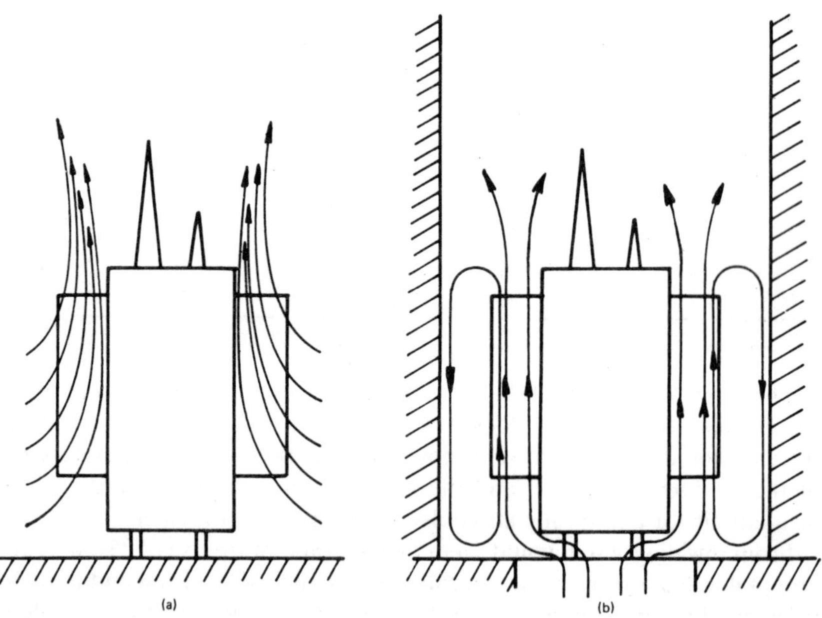

Figure 10.5. Flow of air near a transformer (a) installed outdoors (b) installed indoors

When outside, the velocity of the air increases from the bottom of the transformer and the fresh air approaches horizontally for a large proportion of the height.

When inside, the flow of the air is impeded by the presence of the walls and part of it is recycled by eddy movements. This is equivalent to an increase in the ambient temperature.

For these reasons, the average temperature rise of the volume of air passing through the enclosure is taken as 15° in the calculations and the numerical examples which follow.

Circulating head

The circulating head is the difference in weight of two columns of air of unit area and equal height h, one at the external temperature t_E, the other at the internal temperature t_S.

$$p = 1.293\left(\frac{273}{273 + t_E} - \frac{273}{273 + t_S}\right) \times \frac{H}{760} \times h(\text{kg/m}^2 \text{ or mm water})$$

$$p = \frac{353(t_S - t_E)}{(273 + t_S)(273 + t_E)} \times \frac{H}{760} \times h(\text{kg/m}^2 \text{ or mm water}).$$

Table 10.2 gives the circulating head at 760 mm mercury for a height of 1 m and various external temperatures t_E and temperature differences $t_S - t_E$.

Table 10.2 Circulating head in mm water per metre height

External air temperature t_E (°C)	Temperature rise $t_S - t_E$ (°C)			
	10	15	20	25
−20	0.0531	0.0781	0.1022	0.1255
−10	0.0492	0.0724	0.0949	0.1165
0	0.0457	0.0673	0.0883	0.1085
10	0.0426	0.0628	0.0823	0.1012
20	0.0398	0.0587	0.0770	0.0947
30	0.0372	0.0550	0.0721	0.0888
40	0.0349	0.0516	0.0677	0.0834

The formula and the table correspond to columns of air at constant temperature, which is virtually true for the outside air and the column of hot air above the transformer. At the level of the transformer, the temperature rises from the bottom of the radiators, where it is practically the same as the external air t_E, to the top where it attains its final temperature t_S. The average temperature $t_i = (t_E + t_S)/2$ is used in the formula with the actual height of the radiators. The error compared to precise calculation by integration is less than 1 per cent.

The column of hot air to be taken into account when calculating the total circulating head is therefore the sum of the two components corresponding to the radiators and the space between them and the average level of the outlet openings. The zone below the radiators can be ignored.

Area of the air passage

The circulating head must cause circulation of air within the enclosure at the required flow, by equalling the sum of the velocity head due to the flow and the loss of head due to friction against the walls and the various other obstructions and changes of direction or area.

$$p = P_D + \Sigma P_i$$

The velocity head is obtained by the formula

$$P_D = \frac{V^2}{2g} \times 1.293 \times \frac{273}{273+t} \times \frac{H}{760} \ (\text{kg/m}^2 \text{ or mm of water})$$

where V = velocity of air in m/s, and
$\quad\ g$ = acceleration due to gravity = 9.81 m/s^2.

The loss of head due to friction against the walls is expressed relative to the velocity head. It is proportional to the length and the perimeter of the duct and decreases as the area increases. It is generally of little importance. For this reason it is useless to carry out laborious calculations in fluid mechanics. A sufficiently precise value is given by the formula

$$P_i = 0.007 P_D \times L \times \frac{s}{P} \ (\text{kg/m}^2 \text{ or mm of water})$$

where L = length of duct in m,
$\quad\ s$ = area of duct in m^2, and
$\quad\ P$ = perimeter of duct in m.

In the same way the head losses caused by changes of direction or area are expressed as velocity head by applying a coefficient ζ as given in Table 10.3.

Abrupt changes of area cause loss of head expressed by the coefficient ζ

$$\zeta = \left(\frac{s}{s_1} - 1\right)^2$$

where s and s_1 are the areas in m^2, and the calculation is related to the velocity v_1.

Table 10.3 Individual head losses

Description	Coefficient ζ
Right angle elbow	1.5
90° bend	1
135° change of direction	0.6
Gradual change of direction	0
Grill:	
mesh of 10–20 mm wire 12/10	1
mesh of 20–30 mm wire 14/10	0.75
mesh of more than 30 mm wire 15/10	0.5

The shutters and grills used at the inlet and outlet openings also introduce head losses. The area taken into account is the net area perpendicular to the air flow. The coefficient for grills is given in Table 10.3.

The calculation for a transformer enclosure is made by successive approximations. Starting with initial dimensions for areas and heights, the calculation is carried out in the sequence indicated, and the areas are checked to ensure that the updraught pressure is greater than the sum of the velocity head and the other head losses. If it is not greater, then the areas must be increased and possibly the height increased.

FORCED VENTILATION OF ENCLOSURES

In certain circumstances, such as restricted area and height of the enclosure, or transformers of high power and high losses, natural ventilation may not be sufficient to create the circulation for the required cooling. In these cases, a fan must be installed.

The required air flow has already been calculated. If the fan blows directly into the enclosure, the kinetic energy is lost. The static pressure alone equals the velocity head and the other head losses.

$$P_{\text{stat}} = P_D + \Sigma P_i$$

The power of the drive motor for a fan efficiency η of about 0.2–0.3 is therefore

$$P = \frac{Q \times P_{\text{stat}}}{\eta} \times 10^{-2} \ (\text{kW}).$$

Examples

The enclosure whose section is shown in Figure 10.2 contains a transformer of 800 kVA, 20/5 kV from which the losses are 10 kW.

The required flow of air at the inlet at 760 mm Hg for a maximum ambient temperature of 40°C and temperature rise of 15°C is

$$Q_E = 2.82 \times 10^{-3} \times \frac{273+40}{15} \times 10 = 0.59 \text{ m}^3/\text{s}.$$

At the outlet, because of expansion it becomes

$$Q_S = Q_E \times \frac{273+55}{273+40} = 0.62 \text{ m}^3/\text{s}.$$

The circulating head has two components. One corresponds to the radiator zone when the average temperature is 47.5°C.

$$p_1 = 1.293\left(\frac{273}{273+40} - \frac{273}{273+47.5}\right) \times 2 = 0.053 \text{ mm water.}$$

The other corresponds to the zone between the radiators and the air outlet openings.

$$p_2 = 1.293\left(\frac{273}{273+40} - \frac{273}{273+55}\right) \times 3 = 0.155 \text{ mm water.}$$

Therefore the total circulating head $p = 0.053 + 0.155 = 0.208$ mm water.

The detailed calculation for head loss is carried out as follows:

Losses on entry

area of inlet: $2.6 \times 0.5 = 1.3 \text{ m}^2$,

velocity of air: $\dfrac{0.59}{1.3} = 0.45$ m/s,

loss of head on entry: $\dfrac{(0.45)^2}{2 \times 9.81} \times 1.293 \times \dfrac{273}{313} = 0.012$ mm water

loss of head due to a grill with mesh 15 mm
and wire 1.2 mm ($\zeta = 1$) $\qquad\qquad\qquad = 0.012$ mm water

$\qquad\qquad\qquad\qquad$ Total $\qquad = 0.024$ mm water

Losses in the first elbow

area of duct: $2.6 \times 0.4 = 1.04 \text{ m}^2$,

velocity of air: $\dfrac{0.59}{1.04} = 0.57 \text{ m}^2$,

loss of head in elbow ($\zeta = 1$)

$$\frac{(0.57)^2}{2 \times 9.81} \times 1.293 \times \frac{273}{313} = 0.019 \text{ mm water.}$$

Losses in the horizontal duct

area of duct: 1.04 m^2,

velocity of air: 0.5 m/s,

length of duct: 1.5 m,

perimeter of duct: 6 m,

loss of head in the duct: $0.007 \times 0.019 \times 1.5 \times \dfrac{1.04}{6} \simeq 0.$

Losses in the second elbow

negligible $\simeq 0$

Losses in the cell

negligible due to large area for passage of air $\simeq 0$

Losses on exit

area of outlet: $1.35 \times 0.55 = 0.75$ m^2,

velocity of air: $\dfrac{0.62}{0.75} = 0.83$ m/s,

loss of head on exit: $\dfrac{0.83^2}{2 \times 9.81} \times 1.293 \times \dfrac{273}{328}$ $= 0.038$ mm water

loss due to shutters: $(\zeta = 0.5)$ $= 0.019$ mm water

loss due to grill: $(\zeta = 1)$ $= 0.039$ mm water

————————

0.095 mm water

Total losses: $0.024 + 0.019 + 0.095$ $= 0.138$ mm water

As the circulating head is 0.208 mm water, ventilation of the enclosure will occur.

Supposing that a more powerful transformer having double the losses (20 kW) is installed in the same enclosure.

For the same temperature rise, the circulating head remains the same but the required air flow is doubled to 1.18 m^3/s, and the ventilation losses are quadrupled to 0.552 mm of water. The additional flow must be provided by a fan with the minimum characteristics of

output: $1.18 \times 3600 = 4250$ m^3/h,

static pressure at this output: $0.552 - 0.208 = 0.344$ mm water,

motor power: $\dfrac{1.18 \times 0.344}{0.2} \times 10^{-2} = 0.02$ kW.

VENTILATION OF HIGH POWER TRANSFORMERS

Two types of problems are created by high power transformers.

Installation inside a power station or a substation (e.g. urban transformer station)

In this case the transformer is placed alongside a building wall with open space on the outside. The radiators or air coolers are situated on the outside and are connected to the transformer by pipework passing through the wall. The losses dissipated by the transformer tank are removed by the natural movement of air in the building which is generally of substantial size. In winter it is possible to recover the heat from the air coolers for heating purposes.

Installation outside with soundproofed enclosure

As before, the cooling equipment is placed in open air and connected to the transformer by flexible pipework. The temperature of the air space between the tank and the acoustic enclosure is less than that of the oil. However, in summer it is likely to reach high values depending on the energy loss dissipated by the tank. It is sometimes necessary therefore to install a fan, with silencers on the air inlet and exhaust.

It is not possible to apply general rules for these two cases to determine the size: each installation must be treated individually.

11 Noise limitation

THE PROBLEM OF NOISE

Transformers in service emit a characteristic hum, to which may be added the noise produced by the auxiliaries, pumps and cooling fans.

Over the last 20 years this problem, which hitherto had been of secondary importance, has increasingly become a matter of concern. As a consequence of the increasing consumption of electrical energy, there has been a

Figure 11.1. 100 MVA, 220 kVA transformer in a sealed noise enclosure

need for transformers of increasing power whose noise level has increased correspondingly.

Moreover, the tendency to install transformer stations close to areas of housing or inside industrial workshops for direct supply to processes has made them more environmentally obtrusive.

ORIGIN OF TRANSFORMER NOISE

The basic cause of transformer noise is the magnetostriction of the sheets in the magnetic circuit. Variations in the magnetic induction subjects the sheets to periodic variations in length, the amplitudes of which are in the order of microns per metre length.

The fundamental frequency is double that of the system frequency, 100 Hz, but the phenomenon is not linear as there are also numerous harmonics. Furthermore the various parts of the transformer, starting with the magnetic circuit itself, are liable to vibrate under the excitation effect of the magnetostriction. The noise is transmitted from the magnetic circuit to the tank either by direct conduction to the support points or through the oil and insulating material. The tank and radiators radiate the acoustic energy into the ambient air.

Another source is the vibrations of the magnetic sheets perpendicular to their surface either at the edge of the core packets, or at the joints between leg and yoke.

The windings through which the current flows are also the source of oscillating forces, radial and axial. Their amplitude is very low, and the increase in noise level which results is not detectable. However, in the case of transformers for rectifiers, the sound intensity increases and the frequency spectrum changes because of harmonics in the voltage during the increase from no load to full load.

The cooling fans and pumps emit additional noise.

Finally, vibrations are transmitted to the foundations and they can cause resonance of neighbouring metal structures. Stationary waves can arise in walls that are very close to transformers.

MAGNITUDES OF SOUND

The human ear is more sensitive to the ratio of sound excitation than to its absolute value, and this explains the logarithmic scales adopted for the magnitudes used. It is possible to measure near each noise source the level of sound pressure expressed in decibels (dB) above the reference level, which is the limit of audible sound at 1000 Hz: $p_0 = 2 \cdot 10^{-5}$ pascal.

$$L_p = 20 \log \frac{p}{p_0}$$

where p is the sound pressure in pascals.

Thus decibels do not represent units of magnitude but ratios of magnitude and must be accompanied by a statement to this effect.

In the same way a level of sound power is related to a reference level: $p_0 = 10^{-12}$ watts.

$$L_p = 10 \log \frac{p}{p_0}$$

where p is the sound power level.

The apparatus used to measure sound, the sound level meter, is defined by standards and its reading can be weighted as a function of the frequency in order to approach as near as possible the sensitivity of the human ear, which itself varies with the frequency and sound power. The measurements are made with weightings A and C. The first corresponds to the sensitivity curve of the ear at the 40 dB and 1000 Hz level, while the second gives the actual physical noise. The first enables the nuisance value to be assessed, the second assesses the sound emitter.

METHOD OF MEASUREMENT

In France the measurement of transformer sound is made in accordance with EDF standard HN 52–02. This standard treats the transformer and its auxiliaries separately and takes into account the surface radiating the noise.

In this way, sound sources with different noise spectra and directional radiation can be distinguished, and it is possible to determine the level of sound generated at any point in the environment.

The measurements are made at points equidistant from the prescribed surface, 30 cm from the vertical surface enclosing the transformer and its accessories. An average value is calculated to which is applied the correction for background noise and possibly that due to the enclosure. The weightings A and C are both used. To make it easier to use the results, the sound level is related to a hemispherical surface of 3 m radius or an area of 1 m^2.

For fans and pumps the prescribed surface is of semi-cylindrical or hemispherical shape at 1 m from their centre. The measurements and calculations are carried out in the same way as for the transformer.

VARIATION OF NOISE WITH DIFFERENT PARAMETERS

The principle parameters influencing noise are either external (distance) or internal (frequency, flux density, mass, quality of magnetic sheeting and operation).

According to the laws of acoustics, the volume of sound decreases with the square of the distance d from the assumed point source, i.e. the centre of the equivalent hemisphere.

$$L_{p(d)} = L_{p(3m)} - 20 \log\left(\tfrac{d}{3}\right)$$

where d is in metres.

Thus when the distance is doubled, the sound level decreases by 6 dB.

Other things being equal (mass, induction, etc.) the sound volume varies with the square of the frequency.

$$L_{p(f)} = L_{p(50)} + 20 \log\left(\tfrac{f}{50}\right)$$

The frequency of distribution systems is maintained constant to a fraction of a hertz, and the above equation is given only as a guide.

Magnetostriction varies parabolically with the flux density and also depends on the direction of the lines of flux relative to the direction of rolling of the sheeting. It is minimum when in the same direction. A variation of 10 per cent in the flux density relative to the rated value produces on average a difference of about 3 dB. Figure 11.2 shows the

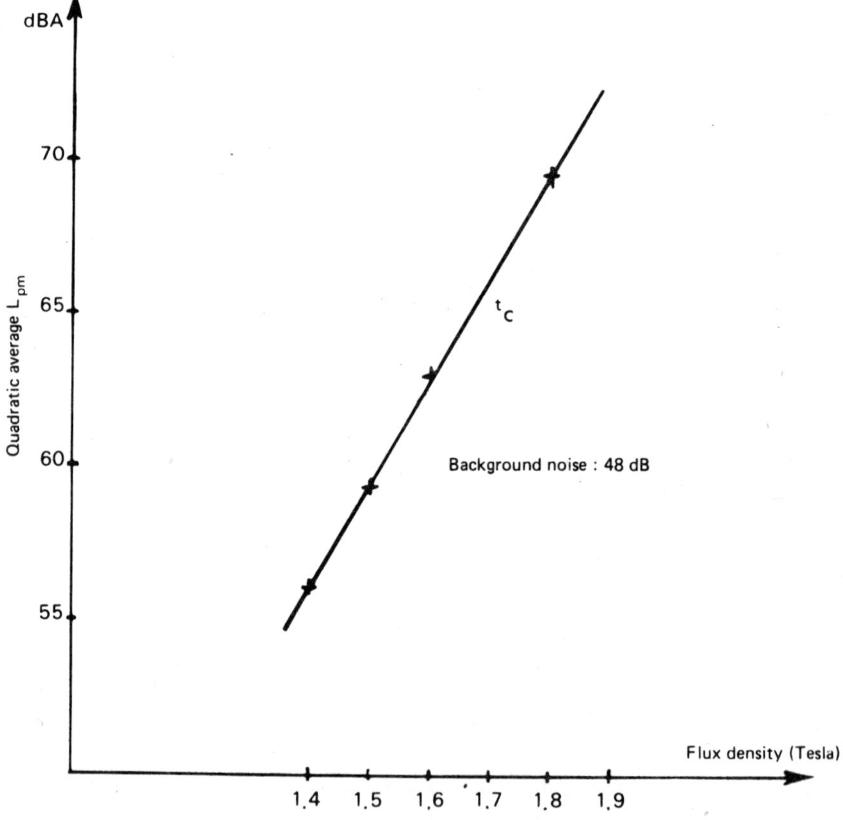

Figure 11.2. Variation of noise level of a 20 MVA transformer

results of measurements carried out on a 20 MVA transformer of recent design, for which there is an almost linear variation of 3.4 dB/0.1 tesla between 1.4 and 1.8 tesla.

The level of magnetostriction being uniform, it can be shown that the sound power varies with the cube of the linear dimensions, i.e. with the volume (or the mass M) of the core, or with the electrical power raised to the power of three-quarters.

$$L_{p(3m)} = L_0 + 20 \log M,$$
$$L_{p(3m)} = L_0' + 15 \log (\text{kVA}).$$

Observations made on a large number of transformers confirm these laws statistically. Some departure from them is due to the levels of magnetostriction being necessarily variable and to possible resonance.

Improvements in magnetic sheeting from the point of view of specific losses have been accompanied by a reduction in the level of magnetostriction. This very sensitive measurement is used by some steel manufacturers to check their manufacturing processes.

During the construction of a core, the flatness of the sheets, the type of joint, the quality of the interleaving at the joints and the clamping, all have some influence on the noise level.

CONSTRUCTIONAL METHODS FOR LIMITING NOISE

The prime consideration is to reduce noise at its source, i.e. the magnetic circuit. Improvements over the years in the construction and manufacture of the magnetic circuit have had a positive effect on the noise level. These include:

> cold rolled grain-oriented plate, with low magnetostriction and improved flatness,
> mitred joints, ensuring a uniform distribution of flux and a reduction of cross flux,
> search for the maximum uniformity in the circuit,
> elimination of clamping bolt holes,
> use of resin impregnated glass-fibre bands instead of core bolts,
> gluing of the core packets.

These measures have reduced not only the losses and the no-load current but also the noise level by 5–10 dB.

A study should also be made of the resonant frequencies of the magnetic circuits, as a result of which certain dimensions conducive to resonances at one of the harmonics of the magnetostriction wave can be avoided.

The above comments relate to normal transformer construction based on optimization of the manufacturing cost plus a factor representing the cost of the losses. When the noise level is one of the principal factors specified by the user, special measures must be taken. Firstly, reduction of the induction by 10 per cent leads to an improvement of about 3 dB. One cannot, therefore, go very far using this method without increasing the size of the transformer and incurring unacceptable costs.

The transmission of sound energy from the magnetic circuit to the tank takes place in two ways: by direct mechanical contact through supports at the bottom of the tank, lateral bracing and top connections, and through the windings, the insulating materials and the oil.

As these two act together, it is necessary to deal with both together to obtain any appreciable result. The ideal solution would be to place dampers at all mechanical connections, and to line the interior of the tank with an absorbent material. The risk of disintegration of any plastic material used, and the possible obstruction of the cooling circuits, makes this solution unacceptable. The use of dampers alone does not provide significant improvement.

As most of the sound energy is transmitted to the tank, this is made as rigid as possible. The general construction now used for tanks makes it possible to employ vacuum treatment. This makes a considerable contribution to reducing noise. The addition of stiffeners reduces the size of the individual vibrating surfaces. Further measures can be adopted, either by using dampers between the stiffeners adjusted for frequencies between 100 and 200 Hz or by completely lining the external surface of the tank with panel dampers. Improvements of the order of 10 dB in the first case and 5 dB in the second can be obtained.

Among the accessories, the fans merit the most attention as they contribute considerable noise to a transformer in service. The remedies consist of reducing the speed of rotation to 750 rev/min, the choice of a small number of blades of appropriate section, the use of elastic support for the fan, and proper aerodynamic location.

Depending on the reduction required in the noise level, some of these measures will be used together, enabling a reduction of 10–15 dB to be achieved. But at this stage, external methods of reducing noise should also be considered and the costs compared.

EXTERNAL METHODS OF SOUND PROOFING

These methods consist either of screens placed close to the transformer on one or more sides, or of complete enclosures.

In the first case, an improvement is sought in a given direction. A wall, usually in concrete, is built as close as possible to the transformer, perpendicular to the direction of the area to be protected. The edges of the screen diffract the sound waves and it must therefore be sufficiently wide and high (see Figure 11.3). The attenuation depends on the difference between the

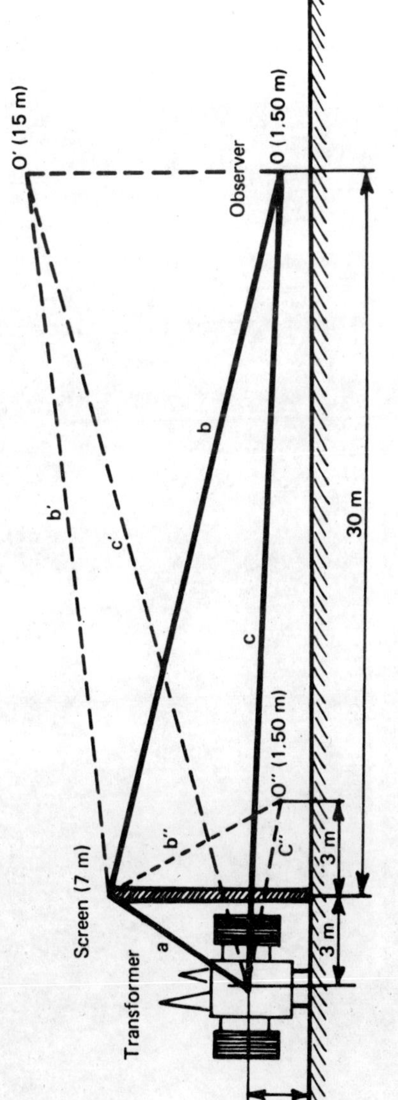

Figure 11.3. Effect of a screen for various observation positions. 0 (12.9 dB); 0' (9.2 dB); 0" (15.5 dB) at 100 Hz

Figure 11.4. 290 MVA, 235 kV transformer with sound-proofed walls

diverted path $(a+b)$ over the screen and the direct path without the screen.

$$\Delta L = 10 \log \left[\frac{20}{\lambda} (a + b - c) \right]$$

where λ is the wavelength of sound in metres.

The calculation must be carried out for the predominant harmonics in the sound spectrum, the first two of which are usually $\lambda = 3.4$ and 1.7 m. To prevent sound escaping around the two ends of the screen, it must extend beyond the transformer at least as much as it extends above.

An attenuation of the order of 10 dB can be expected with this arrangement.

If the area to be protected extends all round the transformer, a screen must be provided on all four sides. The calculation is made in the same way for each screen. The coolers are placed outside the enclosure which must be provided with a sound-proofed access door.

When the required attenuation is as much as 15–20 dB, it is necessary to completely enclose the transformer. The enclosure can be made in block work or in metal (Figure 11.5). The coolers are placed outside on separate plinths and the connecting pipes to the transformer have flexible sleeves.

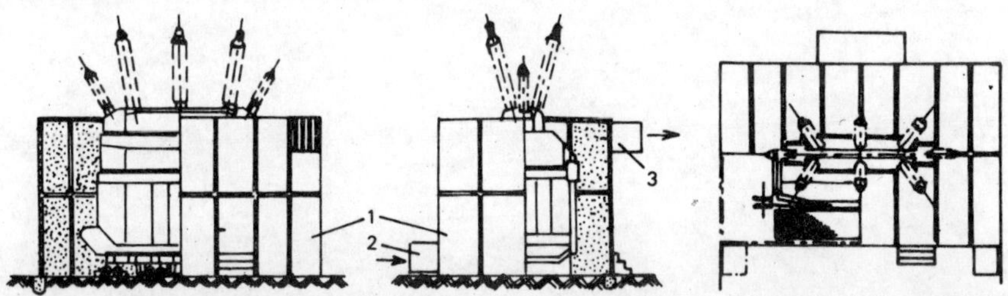

Figure 11.5. Sound-proofing enclosure for a 220 kV transformer. 1. Sound-proofing enclosure, 2. Air inlet silencer, 3. Air exhaust silencer. The cooling installation and the expansion tank are erected separately.

The openings for the bushings are treated in the same way. Ventilation may become necessary if the losses dissipated by the tank are high. Silencers are fitted to the air inlets and outlets, and the fan is of a low noise type.

High performances can be attained by lining the interior surface with an absorbent material or by constructing the wall with material resonating to the frequencies to be eliminated and consisting of hollow blocks of appropriate dimensions.

CONCLUSION

When the noise level of a transformer is such that it might cause a nuisance in the neighbourhood, a reduction can be obtained by measures taken on either the inside or the outside of the transformer. It is possible to reach 10 dB by these two methods. The choice is principally an economic one. Above 15–20 dB, only a complete enclosure can achieve the desired result.

12 Tests

THE OBJECT OF TESTS

As with all industrial electrical equipment, transformers leaving a production line pass through a test laboratory for a final check.

The tests to be carried out on power transformers are always specified in the sales contract and they comply with the appropriate international standards (IEC) or the standards of the country of manufacture or destination.

Special transformers such as those for furnaces or rectifiers, or those required to operate in unusual conditions, or serving as prototype for a series, may have to undergo particular tests. These will be the subject of contractual clauses at the time of placing the order.

The standards meet the following requirements:

define both the tests representative of the service conditions of the transformer, and the quality control tests,
define the measurements to be made, their value and tolerances,
establish the methods of carrying out the tests clearly and precisely,
take into consideration the normal equipment in a test laboratory,
take into consideration the economic cost of the tests relative to the cost of the transformer.

Some of these requirements conflict and compromises have to be made.

Over the years, as experience has been gained, the types of tests have increased and have been perfected. A better understanding of the stresses to which the equipment is subjected under normal or transient conditions

Figure 12.1. Three-phase 600 MVA auto-transformer (400/225/21 kV) during short-circuit tests

and the behaviour under such influences has been made necessary to define these tests.

Industrial development based principally on the use of electrical energy, the increase in the power of interconnected systems, the increase in transmission and distribution voltages and the use of equipment loaded to its full capacity, have all increased the stresses to which transformers are subjected, and the requirements in terms of their reliability.

Thus since the end of the second World War surge impulse, switching, over-voltages, short-circuit and partial discharge tests have all appeared. Heat run and noise tests have been extended to transformers of several hundreds MVA per phase.

TEST FACILITIES

The test facilities include all the equipment for obtaining the necessary voltages, currents, frequencies and powers, as well as the measuring equipment.

At Alsthom, three high voltage laboratories are equipped in this way, two for testing equipment on completion of manufacture and one for research and development. They enable no-load and load performance tests to be carried out as well as dielectric tests.

The largest of the laboratories (65m × 30m × 31m usable height) can accept units up to 600,000 kg weight. Its size makes it possible to test equipment of 400, 500, 800 kV, and equipment for the future systems at 1200 and 1500 kV. The testing facilities include:

equipment for any test on the cooling devices – radiators, air coolers, water coolers,
several large alternators, one of them 27 MVA coupled to a 7000 kW motor,
a bank of capacitors of 140 MVAR with matching transformer,
an impulse generator of 400 kJ at 4000 kV,
a 200/300 Hz alternator rated at 4 MVA coupled to a 1500 kW motor,
a bank of capacitors of 140 MVAR at 1000 kV (phase-earth) which enable reactors to be tested using the principle of series resonance.

In addition, these laboratories are provided with equipment for the most up to date analyses:

measurement and location of partial discharges
digital recording of measurement and symptoms during testing,
location and measurement of hot spots,
analysis of gases dissolved in the oil and diagnosis of results, etc.

The tests prescribed are carried out on completion of manufacture then repeated for the customer's acceptance tests. This is the case for all transformers with power exceeding several MVA. For low power and standard range transformers, acceptance tests are carried out by sampling, when a series is being manufactured, or more often, with the agreement of the client, the transformer is subjected to routine tests only.

Figure 12.2 Transformer and shunt reactor for 800 kV system under test

CLASSIFICATION OF TESTS

Usually the standards divide the tests into three categories.

Routine tests, to be carried out on all transformers without exception, include:

1. Measurement of winding resistance.
2. Verification of the polarity of the windings, the connections between windings and terminals and the transformation ratio.

3. Measurement of load losses and the short-circuit voltages.
4. Measurement of no-load losses and no-load current.
5. Electrical tests at power frequency (by induced voltage, by applied voltage).
6. Test of ability to withstand full-wave impulses (transformers of rated voltage $\geqslant 220$ kV).
7. Check on the level of partial discharges during over-voltages (transformers of rated voltage $\geqslant 220$ kV).
8. Test on ability to withstand switching surge (transformers of rated voltage $\geqslant 220$ kV).

Type tests, intended in principle to prove a new design or a series of transformers, include:

9. Tests of ability to withstand full-wave impulse (transformers of rated voltage <220 kV).
10. Temperature rise test.

Special tests correspond to particular service conditions or to investigations.

11. Measurement of zero phase-sequence impedance.
12. Measurement of noise.
13. Check on the level of radio interference voltage.
14. Test on the ability to withstand chopped impulse wave.
15. Tests of ability to withstand short circuit.

The individual or routine tests are included in the price of the transformer while the special tests are covered by additional charges which take account of the equipment used and the cost of test personnel. For this reason they are only specified for high power transformers or for special service conditions.

The various tests are reviewed here by considering which aspect of service they correspond to, and what conclusion the user can draw from them with regard to the behaviour of his transformer in service. They are divided according to their type:

check on characteristics (connections, ratios, losses),
dielectric tests (at power frequency and impulse),
operating tests (heating, noise, short circuit).

CHECK ON CHARACTERISTICS

This enables the user to verify the correct functioning of the transformer and compliance with the guarantees in accordance with his technical specification.

The arrangement of winding connections determining the vector group and the transformation ratio on all the tappings are usually measured with a special bridge at low voltage. A knowledge of these characteristics is indispensable for parallel operation at any time, whether present or future. The error in measurement is much less than 0.5 per cent. The measurement of the resistances enables the losses due to the I^2R effect only in the windings and connections to be determined. It is carried out using direct current at a fraction of the nominal current. Abnormal results should lead one to suspect defective contact at joints.

For convenience the measurement of load losses is usually carried out by short circuiting the low voltage winding and supplying the high voltage winding, because the required voltages and currents are more easily produced this way.

For these two tests the windings must be at a uniform and well defined temperature, preferably the ambient temperature t. The total losses are equal to the sum of the I^2R losses and the losses due to eddy currents (called 'supplementary') which arise in the windings or the adjoining metallic parts:

$$W_t = W_J + W_F$$

By calculation these values are corrected from the ambient temperature t to the standard specification temperature, 75°C, by the formula:

$$W_{75}{}^\circ = W_J \times \frac{310}{235 + t} + W_F \times \frac{235 + t}{310}$$

This formula shows that the I^2R losses, which are a function of the imposed current, increase with the resistance and the temperature, while the supplementary losses, which are a function of the induced voltage, vary inversely. The supplementary losses are an indication of the quality of design and the possible presence of hot spots in insufficiently divided conductors or in solid metal parts. Normally they are less than 15 per cent of the I^2R losses. However, in the case of transformers of very high current rating or of high short-circuit voltage, this level can be exceeded.

Measured at the same time as the losses due to the load, the short-circuit voltage is the voltage applied to the winding to cause the rated current to flow. This value is expressed as a percentage of the rated voltage of the winding being supplied and is corrected to the temperature of 75°C by the formula

$$E_{75}{}^\circ = \left[E_t^2 - \left(\frac{W_t}{I}\right)^2 + \left(\frac{W_{75}{}^\circ}{I}\right)^2 \right]^{\frac{1}{2}}$$

The measurement of no-load losses and no-load current requires a voltage that is practically sinusoidal, which is obtained when the alternator and test transformer operate at low load. If it is not, the value obtained must be corrected. The voltage is read simultaneously by two voltmeters, one reading the r.m.s. voltage, the other the mean voltage but graduated in r.m.s. values of a sinusoidal voltage having the same mean value. If the voltage applied is sinusoidal the readings are identical, if not the applied voltage is set by the voltmeter reading mean voltage and the actual power is given by:

$$W = \frac{2}{1 + \left(\dfrac{U_{\text{r.m.s.}}}{U_{\text{mean}}}\right)^2} \times W_m$$

where W_m is the power measured, and $U_{\text{r.m.s.}}$ and U_{mean} are the readings of the two voltmeters.

The correction is not valid for transformers having a winding in star without a neutral point connection, and without a winding in delta. Third harmonics appear in the phase voltage but not in the line voltage.

Measurement of power is made by the standard method of two wattmeters.

As well as verifying the contractual values, the measurement of no-load losses during factory tests enables certain, but very rare, manufacturing errors to be detected. These include circulating currents between windings in parallel, due to different numbers of turns or to a closed turn around part or all of the magnetic circuit.

DIELECTRIC TESTS

The insulation of a transformer in service is subjected to stresses resulting from the electric field between the windings and earth, and between windings or parts of a winding.

Several types of voltage can be applied to a transformer (Figure 12.3).

(a) normal voltage at the power frequency, for which the distribution along the windings is linear,
(b) atmospheric over-voltages of short duration, about 100 microseconds, with extremely rapid increase 100 to 1000 kV/µs. In some cases, the over-voltage is chopped by a flashover to earth, causing an even faster variation of voltage. The distribution of voltage within the windings of the transformer is not uniform, with high gradients close to the ends anad possibly a large inductive and capacitive voltage transfer to the LV winding,
(c) over-voltages due to switching operations in the distribution system. Their duration is of the order of milliseconds. The distribution in the windings is uniform.

For all these voltages the amplitude plays an essential part, but if the characteristic element for voltages at 50 Hz is the permanence of the stress with the effect of dielectric loss and of partial discharges, for the lightning

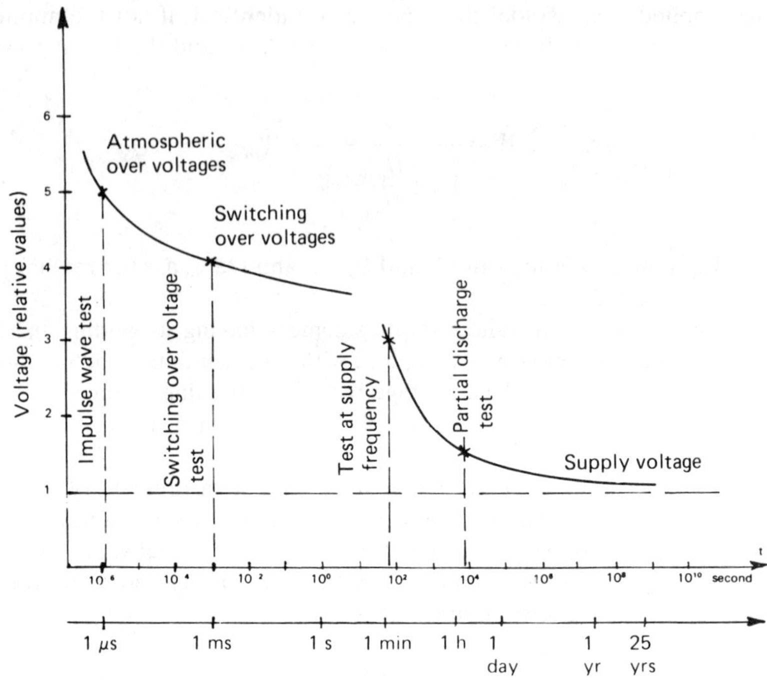

Figure 12.3. Dielectric inflexibility of insulation of a transformer; characteristic voltage/time curve

over-voltages it is the rapid rate of voltage variation and for the switching over-voltages it is their duration.

The dielectric tests must not only simulate the different conditions in a realistic way but also enable a clear and precise diagnosis of the behaviour of the transformer to be made.

The first tests to be standardized were those at power frequency because of their ease of execution. The test by applied voltage is now only used for the low-voltage windings or those connected to a system with isolated neutral or for the neutral point of windings with graded insulation. The whole of the winding tested is brought up to the fixed potential for one minute using a special test transformer, and all the other windings are connected to earth.

The test by induced voltage, the most commonly used, is carried out at increased frequency (e.g. 250 Hz) to avoid saturation of the magnetic core and to reduce the power required. The possible arrangements are shown in the various standards. The duration is fixed at 6000 cycles which in this case is 24 seconds (with a minimum duration of 15 seconds according to the IEC standard).

In transformers with graded insulation, the test by induced voltage is no longer carried out at twice the phase-to-phase voltage but at a voltage

3–3.5 times higher to bring the winding line ends and the terminals to a sufficiently high potential relative to earth (Figure 12.4).

The detection of a fault is relatively easy, based on the behaviour of the measuring apparatus, the relays and possibly the evidence of noise. However, the location of the defect is more difficult, and partial or total dismantling has sometimes been unavoidable. Current methods, particularly those of the analysis and location of partial discharges, help the investigation and in certain cases can avoid large scale dismantling by predicting the onset of a fault at the start of a test before it reaches a

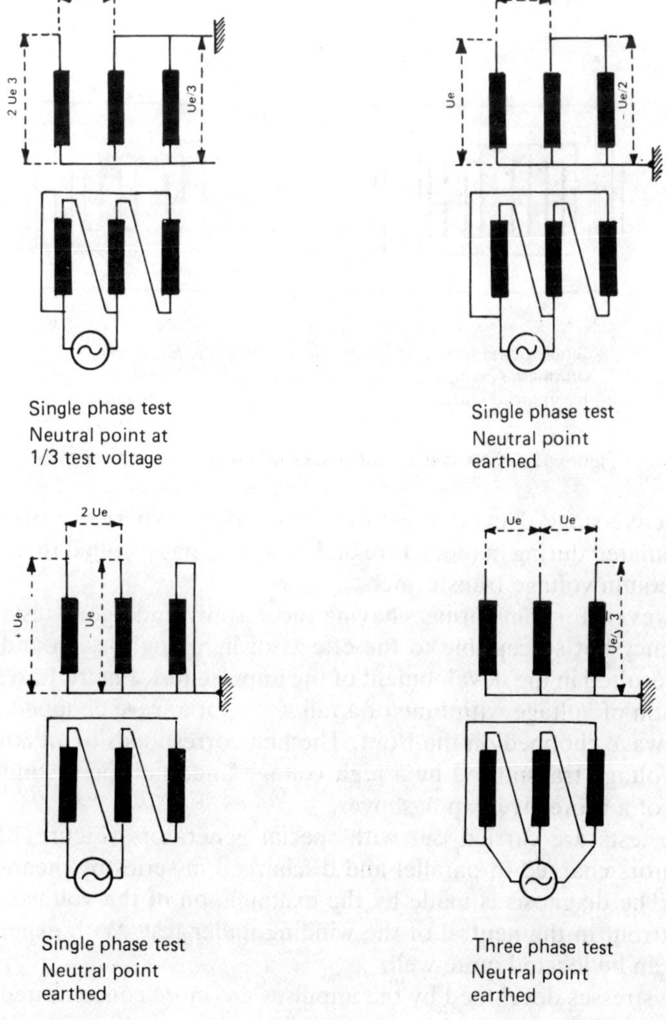

Figure 12.4. Arrangements for induced-voltage test on windings with graded insulation

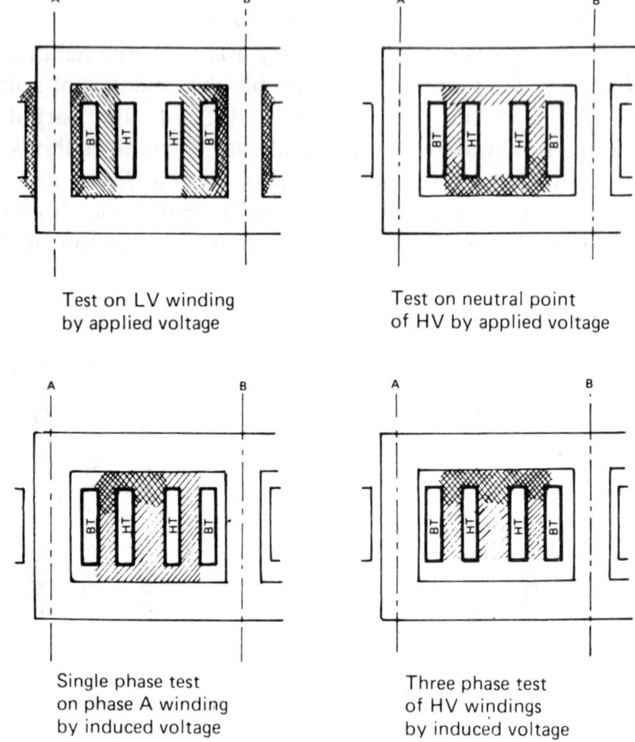

Figure 12.5. Stresses on the insulation during transformer testing

destructive stage. Nevertheless the considerable experience of these tests accumulated during manufacture and in service has justified their retention for medium voltage transformers.

However, for transformers having successfully undergone tests at power frequency but susceptible to the effects of lightning, design and research have resulted in the development of the impulse test. Figure 12.6 shows the variation of voltage with time of a full wave, of a wave chopped after 3 μs and a wave chopped on the front. The first corresponds to an atmospheric over-voltage transmitted by a high voltage line, and the second adds the effect of a protective-gap flashover.

The tests are carried out with special generators (Figure 12.7) using capacitors charged in parallel and discharged in series by means of spark gaps. The diagnosis is made by the examination of the voltage wave and the current in the neutral of the winding under test. With experience the fault can be located quite well.

The stresses developed by the impulses are more concentrated and vary in magnitude in time and in space. The service conditions represented by these tests are of transformers in outdoor stations connected to high

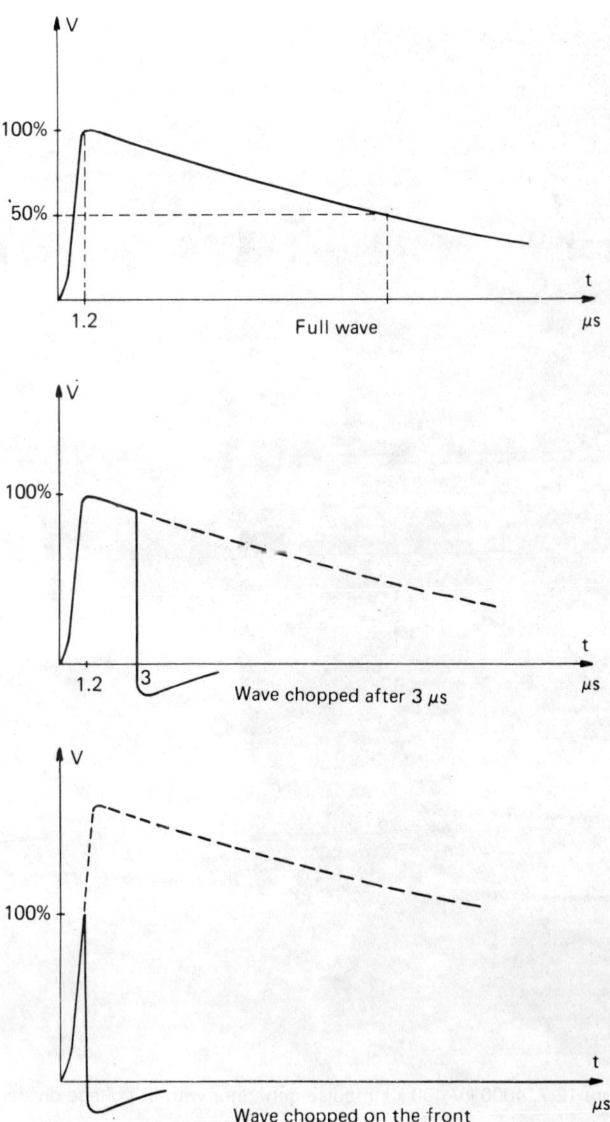

Figure 12.6. Waveforms used in the impulse test

voltage lines. The use of guard wires on the sections of lines close to the substations reduces the risk of nearby lightning strikes. As a result the tests with chopped waves are rarely specified, except for transformers operating in particularly exposed situations.

For high power transformers, and in principle for voltages above 220 kV, the over-voltage test at power frequency is replaced by a switching over-voltage test and a test for partial discharges.

Figure 12.7. 4000 kV 400 kJ impulse generator with its voltage divider

The over-voltage switching test is intermediate between impulse tests (its amplitude is of the order of 80 per cent of the impulse level for the voltage level considered) and the tests at power frequency (the wave corresponds to a half wave at 500 Hz and the distribution of voltage is linear). Its use requires the standard impulse generator to be adapted. In reality, the waveforms of switching over-voltages are very varied, some oscillating, some not. The unidirectional shape chosen is that which subjects the normal insulation to the most severe stress. Model tests have enabled defective insulation arrangements to be eliminated.

The partial discharge test, however, is a quality control test. The transformer is supplied at 1.5 times the rated voltage, and a check is made that the electrical signals generated by the discharges produced in the insulating material are below a predetermined level. This test is considered more thoroughly in the next chapter.

OPERATING TESTS

Several tests concerned with aspects of the operation of transformers under continuous or transient regimes are grouped together under this heading.

The temperature rise test in practice requires almost the same test set-up as for the measurement of losses under load. The sum of the no-load and load losses is first imposed on the windings to obtain the oil temperature, then the losses due to load only to determine the temperature rise of the windings. The measurement is carried out by measuring the variation of winding resistance after connecting the supply. Several service conditions can be simulated, for example, with or without forced ventilation. Checks can be carried out that the guarantees for the maximum temperature rise of the oil and the average temperature rise of the copper have not been exceeded. The values so measured enable the thermal images, if any, to be adjusted.

The test for noise is sometimes necessary when very large transformers are involved or when transformers are installed close to residential areas. The procedure is that described in Chapter 11, Noise Limitation. The test is carried out in the high voltage laboratory where the distances to the walls are large. The noise level is calculated from the measured sound pressures and the sound emitting areas. At the installation site, the attenuation or possible increase in the screening must be taken into account, depending on whether the screens are placed between the observer and the transformer or behind the transformer.

The zero phase-sequence impedance is that which appears between the neutral point and the three terminals of the phase connected together for a winding in star. It is used to calculate earth fault currents in distribution systems. The measurement requires considerable power and, for high power, can be carried out with difficulty under full phase voltage. Because of the losses generated in the tank by the single-phase magnetic flux, the test must be carried out quickly.

Finally, the short-circuit test, together with the impulse test, is that which has shown the most remarkable development. The use of digital computers has enabled manufacturers to determine precisely the electro-dynamic stresses at any point in a winding and to find out the most favourable arrangement.

The availability of high power test facilities provides the means by which the design and construction can be proved. The tests have been carried out on prototypes of transformers of 10, 20 and 36 MVA which have been subjected to 72 consecutive dead short circuits, and then on high power EdF transformers: 600 MVA auto-transformers, nuclear power station transformers of 1080/3 MVA and 1650/3 MVA, auxiliary transformers with more than one secondary and 618/3 MVA transformers for connection to direct current.

CONCLUSION

Tests on transformers are intended to verify that contractual guarantees are being met and to provide the user with the necessary assurance that the behaviour in service will be satisfactory. Their importance and variety increase with the power and voltage.

For distribution transformers up to 4000 kVA of standard design and construction, routine tests are sufficient. Numerous type tests throughout the range determine the choice of standard arrangements.

The situation is more or less the same for transformers of medium power up to 40 MVA. However, particular service conditions should be examined case by case as they may require the addition of an impulse test for transformers over 72.5 kV installed outside, a noise test for transformers installed in a residential area or a heating test for transformers operating under continuous full load.

Finally, impulse tests, switching over-voltage tests and partial discharge tests are usually specified for high power, high voltage transformers.

13 The location of partial discharges

Equipment manufacturers have always tested their products to provide assurance that they will perform suitably in service. When the equipment is mass produced, destructive testing of samples provides an indication of the available safety margins. However, if the equipment manufactured is the only one of its kind or extremely expensive, it cannot be subjected to destructive testing, and frequent tests are carried out during fabrication to provide every possible assurance of achieving valid results. A final inspection test is also performed to confirm that the equipment complies with specified performance characteristics (Figure 13.1).

Partial discharge measurements were introduced to achieve this goal. They were initially performed on high voltage transformers, about 20 years ago, and were the concern of a small number of specialists. Once the utility of these measurements had been demonstrated, they became common practice and maximum discharge levels gradually appeared in specifications.

Large transformer manufacturers were thus confronted with a twofold problem: to check the level of partial discharges and to eliminate the sources of discharges exceeding established limits. Due to the size of the units, the source of partial discharges had to be accurately pinpointed before the equipment was completed.

The General Research Laboratory of the Alsthom Department at Saint-Ouen was assigned responsibility for designing methods and equipment to be used for this purpose.

Two methods were developed: the electrical signals emitted by discharges were analysed to situate a discharge within a specific capacitance zone; ultrasonic pressure waves were detected to locate the source of discharges in an electrical device.

Figure 13.1. Transformer during partial discharge testing

Before explaining these two locating methods, some fundamentals about partial discharges and their measurement are briefly reviewed.

PARTIAL DISCHARGES AND THEIR MEASUREMENT

A partial discharge is a dielectric rupture that occurs in a capacitor directly connected to two electrodes, without having a complete breakdown. The traditional electrical representation of this phenomenon, called the three-capacitor schematic, is shown in Figure 13.2.

Although highly simplified and only partially representative of actual phenomena, this diagram can nevertheless help to explain how partial discharge measurements are performed. The circuit shown is assumed to have no resistance, and capacitance loads are balanced instantaneously. Circulating currents will exhibit the shape of Dirac pulses with a constant frequency spectrum.

Partial discharges thus generate high frequency signals which can be readily filtered from power frequency voltages. A simple computation

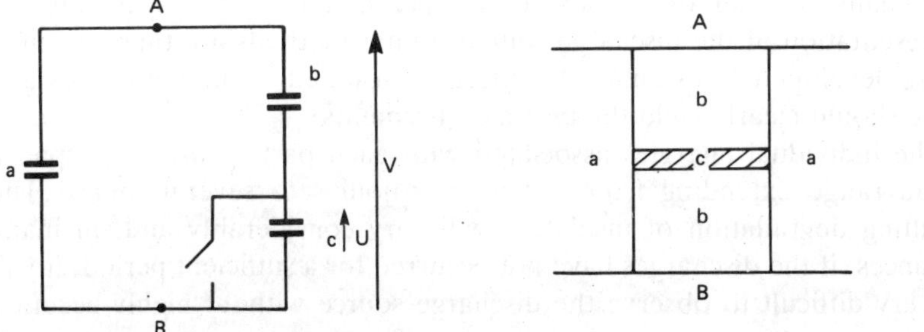

Figure 13.2. Equivalent circuit of a partial discharge. The partial discharge is represented by closing the switch across the terminals of c

shows that, if V is the voltage at the terminals of the device considered, a voltage variation will be created when the switch is closed:

$$\Delta V = -V \frac{b^2}{(a+b)(b+c)}$$

and if b is negligible compared to a and c:

$$\Delta V = -V \frac{b^2}{a \cdot c}.$$

This equation illustrates all the difficulties encountered when a partial discharge is quantified. The voltage variation resulting from the discharge depends not only on the capacitor which is short-circuited (c), but also on the capacitors in series (b) and in parallel (a).

The 'microvolt' measurement defined in NEMA standards corresponds to direct measurement of ΔV. The result depends on the value of capacitor a of the device and the capacitor in series (b). A microvolt assessment is therefore not an intrinsic parameter for characterizing the discharge source.

This is why the apparent charge measurement defined by the IEC was introduced.

For this measurement, a known charge is sent into the device and the resulting voltage variation is measured. By definition, the apparent charge (Q_a) of the discharge is the charge that must be introduced between the terminals to obtain a voltage variation equal to the variation created by the discharge.

A computation based on the elements of Figure 13.2 indicates that the apparent charge Q_a is:

$$Q_a = V_b.$$

This parameter no longer depends on the total capacitance of the device (a), but only on the capacitors in series (b). It therefore depends on the reference terminals selected, hence the name 'apparent charge'.

Actually, neither of the above two parameters provides an intrinsic representation of the discharge source. Other methods are thus currently being developed to evaluate the energy consumed in the discharges and these should clearly yield the best measurements.

The individual energies associated with each partial discharge span a broad range extending from a few microjoules to several joules. The resulting degradation of insulation will vary considerably and, in many instances, if the discharges have not occurred for a sufficient period, it will be very difficult to observe the discharge source without highly accurate location.

Finally, it may also be stated that partial discharge measurements represent today the sole means of estimating the quality of a high-voltage insulation without imposing excessive stresses.

Specifications increasingly include discharge limits, and electrical equipment manufacturers are consequently required to suppress all sources of discharges in excess of these limits.

ELECTRICAL LOCATING METHOD

As explained above, a partial discharge in a capacitor can be evaluated in terms of its apparent charge. This requires two operations:

simulation of phenomena in the capacitor considered,
measurement of the actual discharge.

To obtain a correct result, the discharge must be located in the capacitor in which the simulation is effected. In practice, the test setup often includes several capacitance zones and it is difficult to know in which of these zones the discharge will occur.

Principle

The electrical location method is based on analysis of the electrical signals generated by the discharges and transmitted to measuring impedances. This information is used to determine the capacitance zone where the discharge occurred.

If a simulation is performed in the capacitance zone where the discharge originates, the signals arriving at the measuring impedances will have penetrated the same network and will thus be identical. Otherwise, differences observed will indicate that the simulation has not been conducted in the right capacitance zone.

To illustrate this method, consider the relatively simple case of a test setup with three capacitance zones (Figure 13.3).

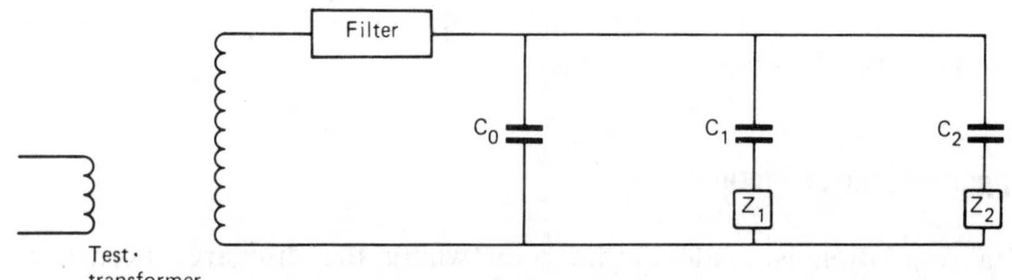

Figure 13.3. Simplified diagram of a test setup

The discharge simulations can be carried out in capacitors C_0, C_1 and C_2; measurements are performed at the terminals of Z_1 and Z_2. Simulation results are given in Table 13.1.

In this case, the discharge can be situated in one or the other capacitance zone simply by calculating the ratio of the two measurements. This method can readily be extrapolated to a more complex arrangement with a larger number of zones. Table 13.1 would be enlarged to include more lines and columns. If the number of measurement points exceeds two, the optimum simulation for the discharge could be determined by one of the methods described below.

Table 13.1

		Measurement		Result
		Z_1	Z_2	Z_1/Z_2
Simulation				
	C_0	80	80	1
	C_1	100	50	2
	C_2	50	100	0.5
Examples of actual discharge in C_1		1000	500	2

Profile method

Simulation in a given capacitance zone will include a series of measurements that can be plotted in the form of a profile (Figure 13.4). The

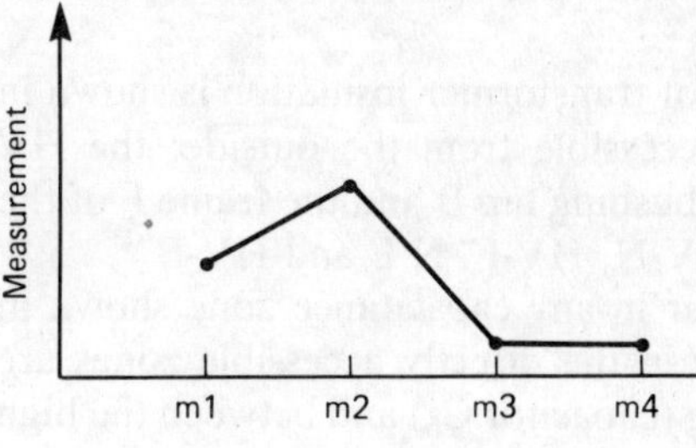

Figure 13.4. Plot of measurements using the profile method

discharge is located by determining the simulation profile that best approximates the discharge profile.

Apparent charge method

If a simulation is made in the zone where the discharge occurs, the apparent charge calculated using the pairs of measurements corresponding to the simulation and to the discharge obtained at the same measurement point should be equal:

$$q_{a\ m,n} = Q \times \frac{M_m}{M_{m,n}}$$

where M_m is the measurement performed at the terminals of impedance Z_m,

$q_{a\ m,n}$ is the apparent charge calculated on the basis of a simulation in capacitor C_n,

Q is the discharge measured at the terminals of Z.

The results are given in Table 13.2, where the charge delivered by the simulator is 100 pC.

Table 13.2

Simulation	Measurement (pC)	
	Z_1	Z_2
C_0	12 500	62 500
C_1	10 000	10 000
C_2	200 000	5 000

The line selected as possibly corresponding to the discharge source is the line with identical values, i.e. C_1.

This example is valid for ideal conditions. In practice, however, measurement errors and spurious capacitance cause some scattering. The most consistent line is therefore selected.

The method is more difficult to apply to transformers because of the complex arrangement inside the tank. Strictly speaking, it cannot be applied unless the discharge occurs in an impedance connected directly to the simulator terminals.

For example, a simplified diagram of transformer insulation is shown in Figure 13.5. Only four points are accessible from the outside: the HV terminal, the neutral terminal N, the bushing tap B and the frame F of the tank. The possible simulations are HV-N, HV-F, N-F and HV-B.

The discharges may of course occur in any capacitance zone shown in Figure 13.5. Only those which occur in the directly accessible zones are correctly simulated, i.e. in the bushing (capacitor C_b) and between the high voltage and frame points (capacitor C_0).

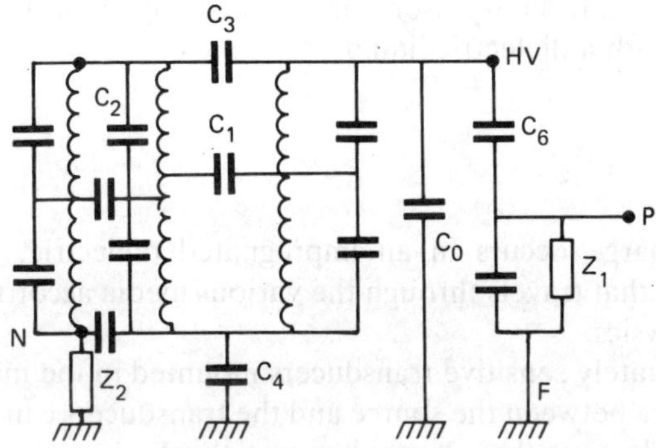

Figure 13.5. Equivalent circuit of transformer insulation

Discharges that occur in C_1, C_2, C_3 cannot be simulated correctly since they are not directly connected to the accessible points because of C_4.

If C_4 is not predominant, the method will apply and the discharges will be attributed to a capacitance between the high voltage and the neutral. To ensure greater accuracy, a discharge at the terminals of C_1, C_2 or C_3 would have to be simulated. This is not feasible using a conventional electronic simulator, which delivers a known charge, but may be accomplished by computer simulation provided an equivalent circuit representing the isolation from measurement frequencies used is feasible. This type of simulation is currently under study.

The method applied to transformers may not seem very accurate, but the results obtained are nevertheless quite useful and often irreplaceable.

In some cases, only the phase on which the discharges are produced can be identified. At other times, more accurate locating is accomplished and this permits the identification of the incriminated coil or of a specific capacitance such as the line insulation or the bushing.

Electrical locating is required to estimate the apparent charge of a discharge and its energy since the discharge cannot be quantified unless the capacitance in which it occurs is known. In the near future, as a result of existing high-performance electronic components, this information will be provided in real time mode. Test verification devices based on this method are moreover under study.

The accuracy of the method nevertheless remains insufficient to enable suppression of a partial discharge source in a large transformer. This is why an ultrasonic locating method has been developed.

ULTRASONIC LOCATING METHOD

The electrical locating method described above is not specific to a particular type of equipment. It can be applied to generators as well as circuit-breakers, models, bushings or transformers. Owing to the detection principle employed, the ultrasonic method is only applicable to insulation impregnated with a dielectric liquid.

Principle

When a discharge occurs in an impregnated dielectric, it generates a pressure wave that travels through the various media according to the laws of classical physics.

Using adequately sensitive transducers mounted in the immediate vicinity, the distance between the source and the transducers can be determined by measuring the time that elapses between the discharge and arrival of the wave at a transducer. A triangulation method is then used to pinpoint the source.

As the discharge time is very short, the pressure wave emitted will include high frequency components well above the audible range. The locating method is therefore referred to as an ultrasonic method.

A locating operation includes two main phases:

signal acquisition,
analysis of results by determining signal transit times and pinpointing the discharge.

Signal acquisition

The instrumentation system used is shown in Figure 13.6. The main component is the transducer, which was designed to provide adequate sensitivity as well as reliability and ease of operation.

The design parameters for the transducers and instrumentation system were determined by analysing travel of a pressure wave from its emission point. This analysis is reviewed below.

At emission, the main characteristics of the pressure wave are its amplitude and frequency. These parameters cannot be modified, and the instrumentation system must be tailored to the characteristics of the pressure wave.

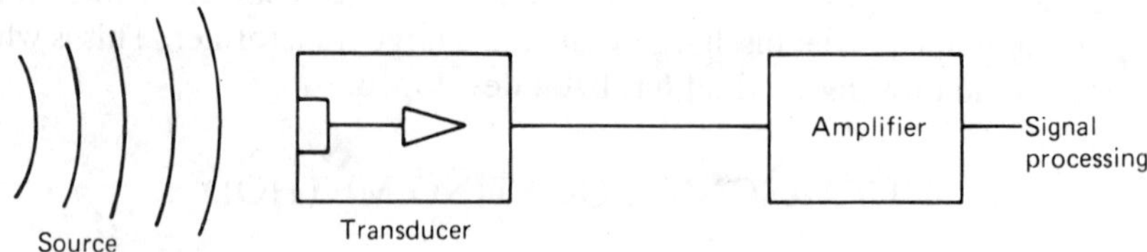

Figure 13.6. Ultrasonic instrumentation system

Table 13.3

	Propagation velocity (m/s)	Relative attenuation compared with oil (dB/cm)
Oil	1400	0
Oil-impregnated paper	1420	0.6
Oil-impregnated board	2300	4.5
Sheet steel	5050	13
Copper	3580	9

The frequencies encountered on models and transformers are so diverse that a broad-band instrumentation system was selected. This led to an increase in electronic background noise, but it is preferable to lose in average sensitivity than to lose sensitivity completely in certain instances.

The wave travels through various liquid and solid media according to classical laws of physics. Each medium penetrated, whether liquid or solid, distorts the wave to some extent.

The principle wave propagation characteristics are given in Table 13.3.

Logically, to achieve maximum sensitivity, the transducers should be placed inside the tank as close as possible to the source.

Since the position of the source is unknown by definition and the transducers cannot be installed easily and safely anywhere inside the tank,

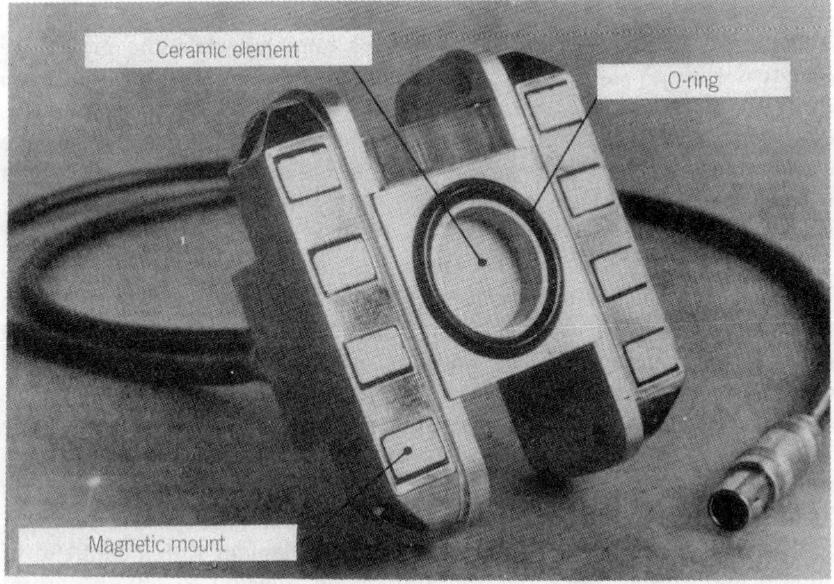

Figure 13.7. Front view of a transducer

Figure 13.8. Transducer mounted on tank wall and being filled with an oil syringe

it appeared preferable to opt for external transducers despite the attenuation caused by the tank wall. The pressure wave should thus be transmitted to the transducer through the tank wall.

As in an optical system, attenuation is minimal when the media are identical on both sides. The transducer is thus designed with oil between the tank wall and the sensing element (Figures 13.7 and 13.8). This considerably augments reliability since, once the oil is in place, a suitable connection is assured. It is moreover preferable to handle some oil, with related difficulties, than to risk performing tests with an inoperative transducer.

The wave thus reaches the sensing element in the instrumentation system, i.e. the piezoelectric ceramic element, which converts the pressure wave into an electrical signal. The dimensions of the ceramic sensing element were determined as a result of theoretical and experimental considerations for maximum efficiency.

The ceramic element characteristics are specified so that transducer sensitivity increases with size. However, for this to be effective, the pressure wave must reach all points of the element at the same time. This limits the diameter of the element to a few centimetres because of the input signal frequencies (10–100 kHz). A transducer element with a diameter of 30 mm and 10 mm thick has therefore been selected to obtain natural frequencies of approximately 20 and 80 kHz.

Signal transmission

At the end of the instrumentation system, the electrical signal is a picture of the pressure wave emitted by the discharge after travel through the transformer insulation. This signal requires processing to provide information enabling location of the discharge.

A transducer element delivers an electrical signal that must be suitably amplified and transmitted. Consequently, a preamplifier was placed in the same enclosure as the transducer element to raise the signal level enough to minimize the effect of cable noise. The input stage of the preamplifier was, moreover, built with components selected for their low-noise characteristics.

Signal processing

As the energy associated with partial discharges is very low, the resulting signal is often quite close to the background level and thus difficult to exploit. The signal-to-noise (S/N) ratio is therefore increased by using an averaging device.

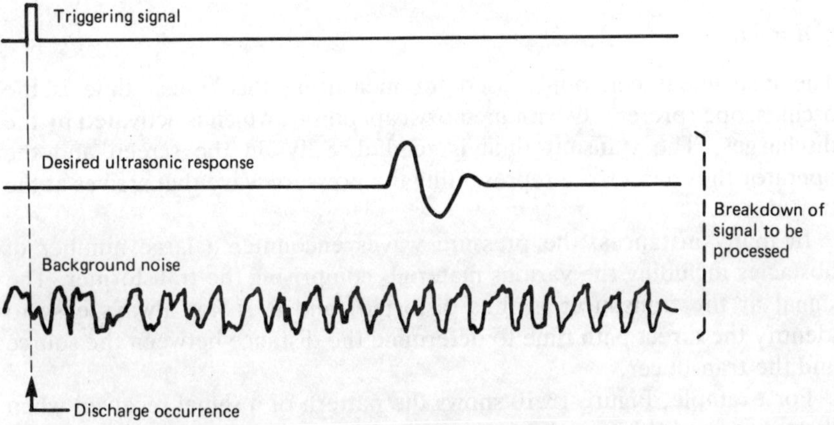

Figure 13.9. Ultrasonic signal and background noise

The averager extracts a reiterative signal from background noise, provided a synchronous triggering impulse is associated with the target signal (Figure 13.9). From this impulse the averager samples the signal to be processed at specified constant intervals throughout the operation. Samples corresponding to delay times given with respect to the triggering impulse are added in a series of memories.

At the end of this operation, it is assumed that n triggering impulses have been taken into consideration; the content of a memory may be considered as the sum of n samples of a background noise and n samples of the target signal. It can be shown that the background input increases with the square root of n while the input due to the signal inreases proportionately with n. This gives an improved S/N ratio with the square root of n.

The requirements for use of an averager are entirely satisfied in the case of partial discharges and attendant ultrasonic responses. Practical experience has shown that enhancement ratios of 100–300 can be obtained by using an averager, limited more by the required application time than the method itself.

To achieve a gain of 100, the operation lasts two or three minutes and a gain of 300 requires over 15 minutes per transducer.

Locating the source and determining transducer-to-source distance

The general method is to measure the time interval t between occurrence of the discharge, indicated by the electrical signal at the terminals of the measuring impedance, and the time of arrival of the pressure wave after travel through the ambient environment at velocity v. The distance is then calculated as follows:

$$d = vt.$$

The instrument commonly used for measuring this transit time is the oscilloscope (preferably with memory capability), which is activated by the discharges. The transmit time is read directly on the screen and the operator thus has a trace representing the pressure wave that arrives at the transducer.

In many instances, the pressure waves encounter a large number of obstacles including the various materials comprising the transformer. The signal at the transducer is thus distorted, and it is not always easy to identify the direct path time to determine the distance between the source and the transducer.

For example, Figure 13.10 shows the pattern of a signal received when the source and the transducer are interconnected by multiple paths. With the propagation velocities in the various materials and the diverse attenuations, the signals range from time t_2, the shortest path, to T corresponding to fadeout, through t_1 corresponding to the most intense direct path to be used for calculating the distance between the source and the transducer. This is a classical example which involves no problems for an experienced operator.

In reality, there are complications because the signals cannot be interpreted properly unless the location of the source is known, and if the location of the source is known, locating is unnecessary. An iterative

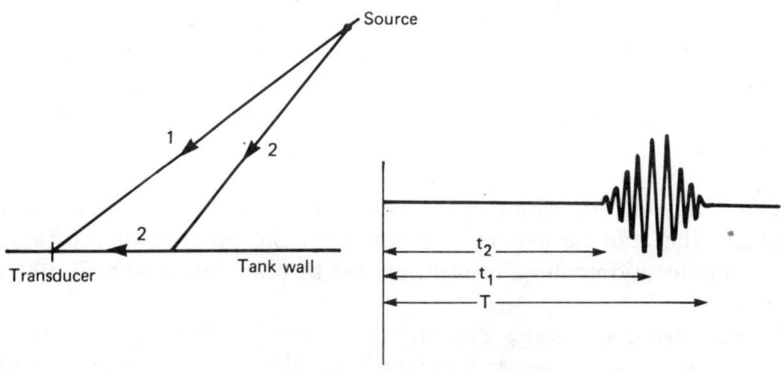

Figure 13.10. Typical compound ultrasonic signal

process is therefore required. In any event, if a response can be detected on one transducer, it should be possible eventually to locate the source.

To reduce the time required and avoid errors, a computer program has been developed.

Computer triangulation program

A locating operation is generally performed in steps. The first step is to determine the section of the transformer where the source is located. The second step is to place a maximum number of transducers in the postulated area and record the responses and coordinates of the transducers in relation to a reference point. The operator then identifies on the recordings the times that appear to correspond to the direct paths through the oil.

A computer provides considerable help since it is not easy to determine the intersection of three spheres. Calculation becomes possible of all intersections corresponding to all possible groups of three transducers (triads), if necessary. This leads to considerable scattering of the points representing a solution because the intersection of three spheres gives two solutions, i.e. points that are symmetrical with respect to the plane passing through the centres, and also accuracy of the solution point and distances measured is sometimes poor for the arrangement of the transducers (e.g. transducers practically aligned).

The data processing philosophy used is based on an initial discrimination. This includes:

1. Elimination of all transducer triads with configurations or measurements that yield inaccurate solutions.
2. Suppression of all triads yielding two solutions both in the transformer tank (selection of correct solution unfeasible).

3. Calculation of coordinate averages to propose a solution.
4. Using the proposed solution, calculation of the corresponding transit time and comparison with time actually measured, plus review in case of a discrepancy, and possible recalculation if the times selected are not correct.

This ultrasonic locating method has been in use for several years. The accuracy that can be expected depends on the position of the fault in relation to locations where transducers can be mounted, within 50–100 mm at best.

If the partial discharge measurements are significantly disturbed by noise, a source can be located without using the electrical signals from the partial discharges, provided the ultrasonic response level is high enough. Instead of considering that the desired solution is at the intersection of the three spheres, where the points are at a constant distance from three fixed points, the solution is considered to be at the intersection of three sheets of hyperboloids of revolution representing the points whose difference in distance from two fixed points is constant.

Once again, the use of a computer is necessary. Another possible application of ultrasonic location is during lightning impulse and over-voltage testing. In these cases, the transducers used must be electrically isolated from the ground (e.g. by fibreoptic links). The signals are obviously no longer repetitive and use of an averager is impossible. However, there are no sensitivity problems since, for the tests, the discharge energy levels are much higher than at power frequency.

CONCLUSION

Both methods described are fully operational and have been used to enhance significantly the state of the art in partial discharge measurement.

The equipment required, installed at a permanent facility (Figure 13.11), permits measurement, interpretation and locating of a partial discharge source with good accuracy within a short time. This enables rapid, effective decision making and maintenance. Applications of the methods described are myriad.

Electrical locating is part of actual measurement techniques, and ultra-sonic locating has made it possible to pinpoint location discharges ranging from 10 picocoulombs to breakdown with an accuracy of 10–50 centimetres.

Locating techniques provide a wealth of information which appears so normal today that there is a tendency to overlook the past. A look at the situation 20 years ago, when these methods did not exist or were only developing, is sufficient to grasp the import of these developments.

Figure 13.11. Partial discharge instrumentation and locating system

Many tests have been performed on transformers equipped with resin-impregnated paper bushings without knowing that the discharges measured originated in the bushings. Today, with the electrical locating method, this information is obtained immediately.

When discharges were detected, many tests and experiments had to be performed before ascertaining that the discharges were located in the transformer. Many will remember external 'discharge hunts', individual installation tests, generator replacements and even moving the tested unit to overcome environmental constraints.

Today, as soon as transducer response is detected, there is no longer any doubt about the possibility of discharges outside the transformer. When no ultrasonic response is obtained, the electrical locating method can be used to trace the source quickly to the generator, cable or environment.

If these locating methods had not been developed, partial discharge measurements probably would not have become so important and transformers would not have benefited from the current substantial increase in reliability.

14 Maintenance

NECESSITY FOR MAINTENANCE

The security of supply from a transformer depends not only on the care with which a manufacturer has designed and constructed it, but also on the steps taken by the user to carry out regular checks on its working condition.

Figure 14.1. Private 220 kV substation equipped with two 80 MVA transformers

The object of maintenance is to ensure correct operation of the transformer permanently over its design life (20–25 years), or even longer if this is justified economically and technically.

Upon installation of the transformer, all steps must be taken to ensure that it functions normally, particularly with regard to the cooling, the auxiliaries and the protective devices. Moreover, the layout of the substation must allow sufficient space for handling and installing the transformer while maintaining the required safe distances from live parts for personnel to work.

Maintenance can avoid or reduce damage. The diversity and scope of maintenance operations vary with the type of transformer, its power and its relative importance in the supply system. Each user develops his own routines, appropriate to his needs. A certain number of general guidelines can be given, but the ways in which they are applied are subject to wide variation.

CHANGES IN A TRANSFORMER DURING SERVICE

Many external factors can be responsible for changes in a transformer over a period of time and may reduce the security of supply. The most important are the actions of air, oxygen and humidity on the insulating materials.

In standard transformers, oil in contact with air absorbs oxygen which diffuses progressively throughout the tank. The humidity results either from the thermal degradation of the cellulose based insulating materials, or from diffusion from the atmosphere. There are several effective protective measures to reduce the harmful influence of the external air. In increasing order of effectiveness these are conservator and air dryer, sealed conservator with gas cushion or membrane.

Oxidation of the oil increases with temperature and the presence of catalysts. The result is the formation of acids and polar products causing an increase in the dielectric power factor (tan δ).

The formation of deposits or sludge with modern oils has become unusual. It can occur however from the mixing of incompatible oils. The presence of water in the oil reduces its dielectric strength, particularly when the water is in suspension or adsorbed onto solid particles.

Humidity absorbed by the insulating materials considerably reduces their dielectric rigidity. The distribution of humidity between the oil and the cellulose based insulating materials varies with temperature. When the temperature increases, the oil dissolves more water and the paper dries. The water contents of paper and oil are respectively 0.5 per cent and 10 ppm on completion of manufacture, and can increase to 1 per cent and 50 ppm in service. The variation is detectable therefore in the oil, and it is a good indication of the state of the insulation.

The prolonged action of heat on the cellulose based insulating materials causes a chemical change, a rupture of the molecular chains (depolymerization) aggravated by the presence of humidity which greatly reduces their mechanical strength. Their resistance to forces during short circuit are thus reduced, without their dielectric characteristics apparently being modified.

The transformer's auxiliary equipment, such as on-load tap changers, fans, pumps and various relays, are also subject to change with time.

Corrosion due to industrial atmospheres (smoke and dust) or marine atmosphere (salt laden mist) has also to be considered. Inclement weather in winter (rain, snow, ice or large temperature variations) subjects protective coatings to greater attack.

INSPECTION OF TRANSFORMERS

Depending on the transformer power and its load duty, inspection periods can be daily or weekly. Their object is to verify that nothing abnormal has arisen with regard to noise, leakages, level and temperature of the dielectric, and operation of the cooling system.

The temperature of the dielectric must not exceed 60°C above the ambient temperature, except when under special load conditions.

CHECKS

A certain number of maintenance operations are carried out periodically during the year taking advantage of interruptions to service, or while in service if the safety regulations permit. Their frequency is determined in each case by experience as a function of the operating conditions and the climate.

Monthly

The air dryer must be inspected. The silica gel must be replaced as soon as one third of its volume has changed colour. The regeneration is carried out in an oven maintained at 100°C for several hours, until the initial colour, usually blue, has been restored. The tightness of the joint and possibly the air breather oil bath must be checked.

A check of the oil-tightness of the transformer may need to be followed by a tightening of the joints showing leakage. The temperature of the dielectric and its level in the expansion tank, which vary with the ambient temperature and the load, must be checked.

The noise from the transformer and its auxiliaries, pumps, fans, must be checked to detect any possible anomalies.

Annually

The insulators must be cleaned.

The tightness of the connections to the terminals must be checked to avoid any abnormal heating.

A check on the functioning of the Buchholz relay can be carried out again. This was tested during commissioning by operating the drain valve, having first closed the valve to the conservator. The alarm and trip contacts must operate successively and not reset until the relay is filled with oil again. The transmission of signals to the circuit-breaker and to the control panel should be checked.

The on-load tap changer is subject to special instructions from the manufacturer. If the operating frequency is high, the number of operations of the tap changer between changes of the contacts may alter the frequency of maintenance.

The following maintenance work must be carried out on the auxiliaries: checks on electrical circuits, operation of measuring instruments (temperature, level) and indicators (thermostats, oil circulation), fire protection and all other instruments for protection, cleaning of control cabinets.

If the coolers are choked up, hot water with detergent can be used for cleaning, or they can be sprayed with a detergent solution and then rinsed with a water jet, or even cleaned with a brush and jet of compressed air depending on what is available and how badly choked they are.

A new coat of paint may be thought necessary, depending on the environment. Painting should in any case be carried out after ten years.

MEASUREMENTS

These are mainly concerned with the dielectric liquid, which should be checked periodically. The sampling and test conditions are described in detail in the various references, and these also set the minimum and maximum values, and the frequency of testing.

The checks are carried out primarily on the dielectric rigidity, the colour and the acidity. The other characteristics such as water content, resistivity, and dielectric loss can be laboratory checked as part of an investigation.

The dielectric in hermetically sealed transformers is not in contact with the exterior and does not need checking for ten years except in case of an incident. The sampling for all transformers immersed in oil should be carried out from the bottom of the tank.

A new method of investigation, now well developed after several years of laboratory work, is the chromatographic analysis of gas, which is either dissolved in the oil or collected from the Buchholz relay. Systematic tests have shown that defects resulting from local heating or partial discharges, which affect solid or liquid materials or both together, cause the production of gases, the principal being H_2, CH_4, C_2H_2, C_2H_4, C_2H_6, CO and CO_2.

The types of gases present in a sample and their relative amounts enable the nature of the fault to be diagnosed whether it is thermal or electrical, and the type of material producing the gas. Tests carried out over a period give some idea of how the defect is developing and indicate when some action must be taken. This relatively expensive test is used systematically in many countries as a preventive measure for high power transformers (greater than 100 MVA). For medium power transformers (5–100 MVA) the method gives useful information when alarm situations arise or damage occurs. It is only used on transformers below 5 MVA in very exceptional circumstances.

It is necessary to be very prudent in the diagnosis when using this method. No absolute rule or limiting value can be given. The interpretation of the results requires a knowledge of the design and the technology of the equipment and its conditions of use.

DIELECTRIC TREATMENT

When measurement of the characteristics of the oil shows values exceeding the limits given in the standards, two situations can arise:

1. If the water content and the electrical characteristics are in question, the oil must be filtered, dried and degassed. The special equipment such as filter, oil and vacuum pumps, degassing tank, reheater and measuring equipment can be installed in a special trailer-workshop allowing treatment to be carried out on-site. The oil is treated in closed circuit until the correct values for the dielectric strength are obtained, this being taken as a guarantee of the quality of treatment.
2. The neutralization index and the surface tension show that the oil is old. Regeneration would be necessary but is not always economical. The replacement of the oil, accompanied by cleaning of the core and windings, is preferable. The old oil can be returned to a refinery or added in small quantities to heating oil.

FAULT TRACING

A fault involving the transformer itself is usually signalled by the Buchholz relay or the earth-leakage protection, before or at the same time as the current differential or over-current protection. The operation of the first two devices, in accordance with the specific instructions given by the manufacturer, is considered below.

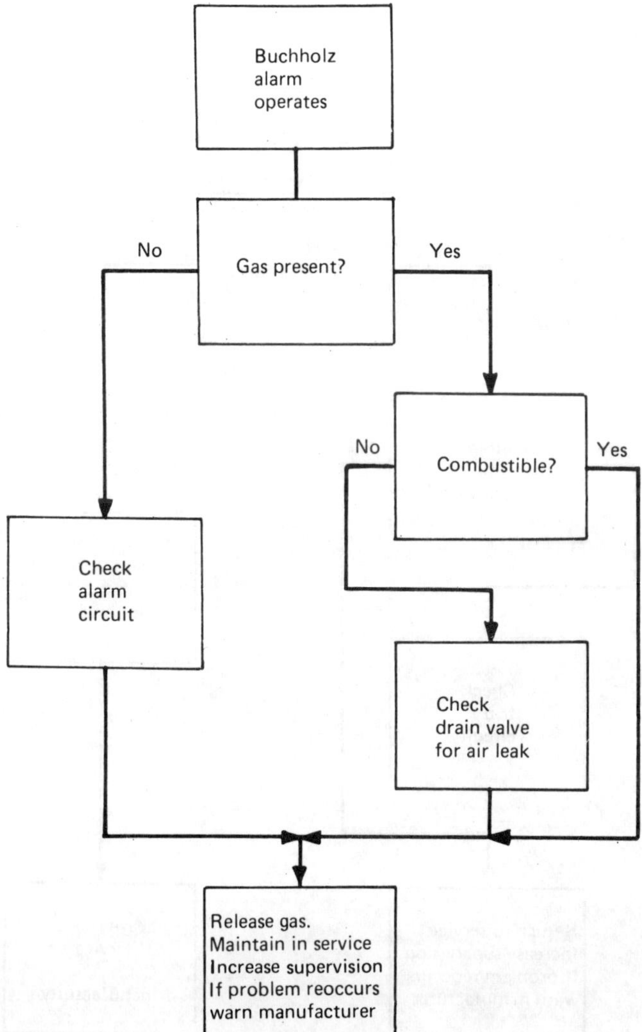

Figure 14.2. Instructions if Buchholz alarm operates

The Buchholz relay can cause the transformer to be switched out in cases of rapid internal pressure rise or production of large quantities of gas, but it also emits an alarm signal when there is a fault with slow production of gas.

Figures 14.2 and 14.3 show the sequence of checks and operations to carry out, and the measures to be taken if the relay should operate. Firstly, check the gases collected, then take a sample for analysis while the transformer is kept out of service, or find the causes of the problem before returning it to service.

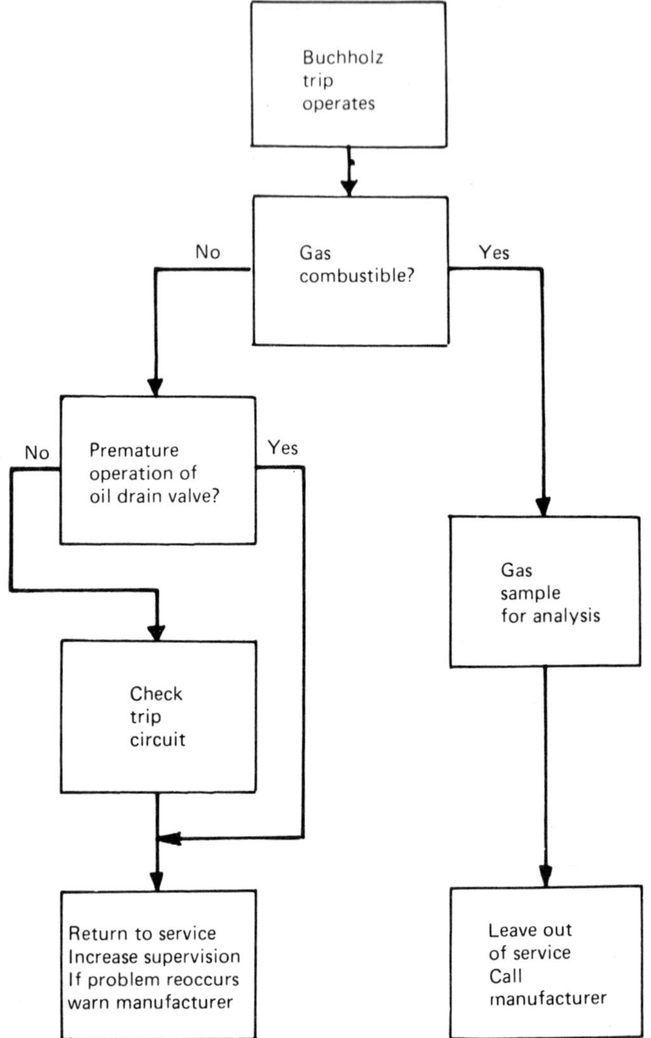

Figure 14.3. Instructions if Buchholz trip operates

The earth-leakage protection (Figure 14.4) can also react to a fault in the insulation of the auxiliaries connected electrically to the tank. After the defective equipment has been removed, the transformer can be returned to service.

The instrument checking the oil temperature is either a thermometer or a thermostat with alarm and/or trip contacts, the latter usually being adjusted to operate 10°C above the alarm temperature. As before, it is necessary to find the cause of the problem and check the adjustment and

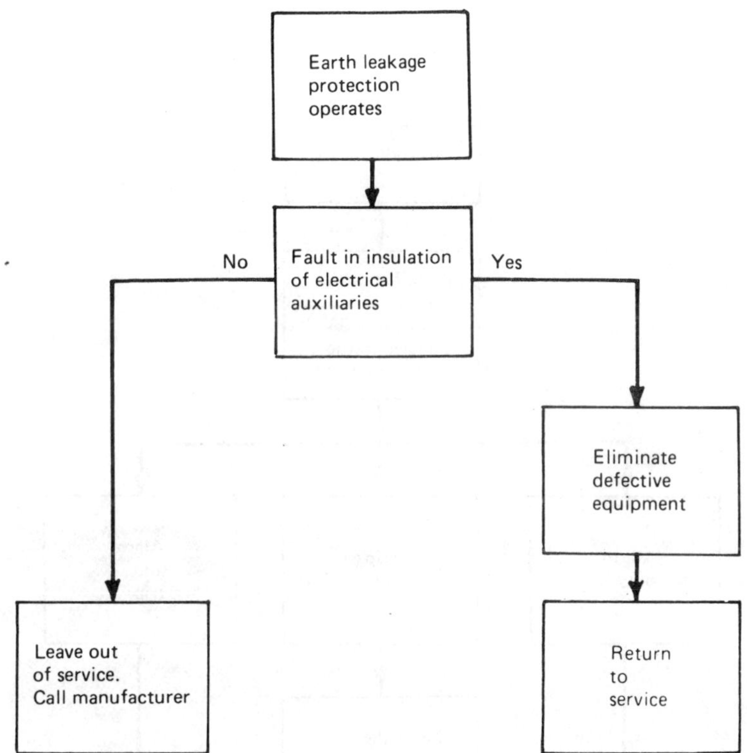

Figure 14.4. Instructions if earth-leakage protection operates

proper functioning of the instrument (Figure 14.5). If all is in order the transformer can be returned to service accompanied by a load reduction to ensure that the oil temperature does not rise above the prescribed limit (see Chapter 6, Overloads).

Where transformers have forced cooling (Figure 14.6), a total or partial stoppage of the cooling system can be shown either by a 'fault' signal on the oil flow indicator, if a pump is involved, or by operation of the thermostats. The remaining cooling capacity may allow the transformer to operate at reduced load. In case of an incident the power is determined by the specific instructions provided for each large transformer. As a general rule, any cooler becomes ineffective if the pump supplying it stops, while a fan stoppage reduces the capacity of the group of coolers to which it belongs. The total stoppage of the cooling system for a period varying from 15 min to 1 h, depending on the transformer and the load conditions before the stoppage, will require the transformer to be taken out of service. The tank alone is unable to dissipate the heat losses even in a no-load situation.

To make fault diagnosis easier, as much information as possible should be obtained, such as conditions in the systems on both sides of the

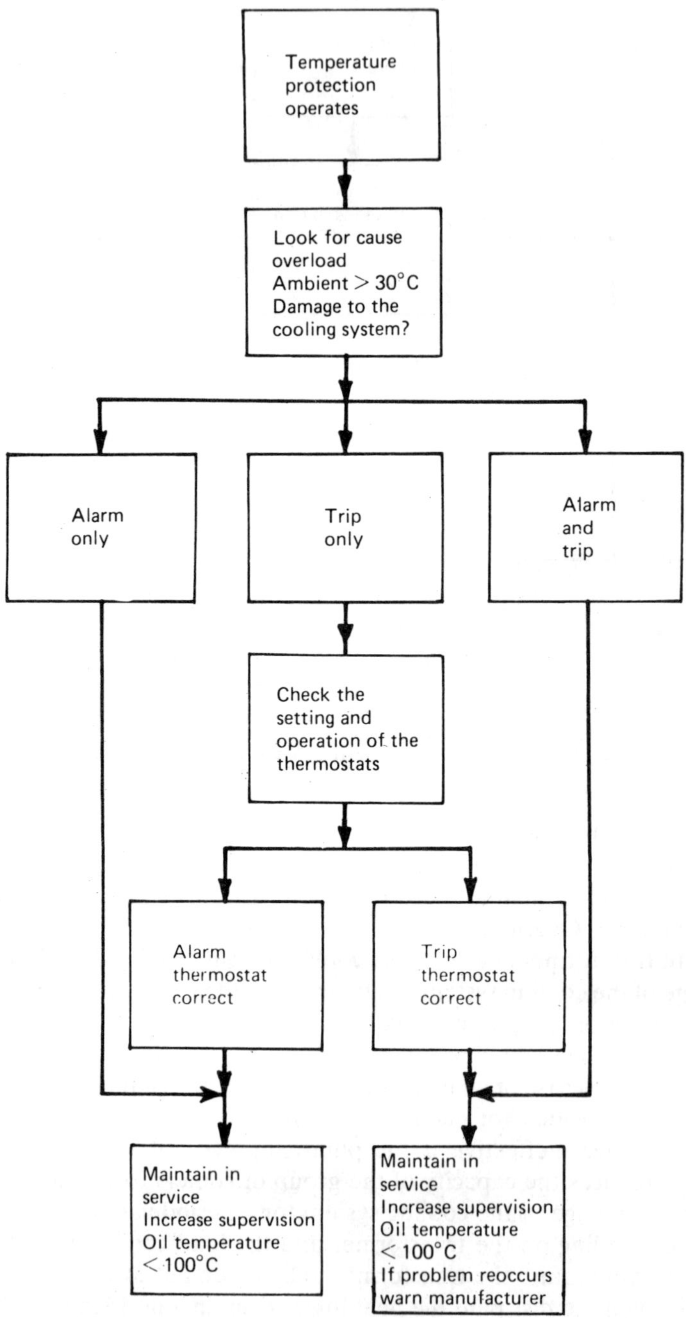

Figure 14.5. Instructions if temperature protection operates

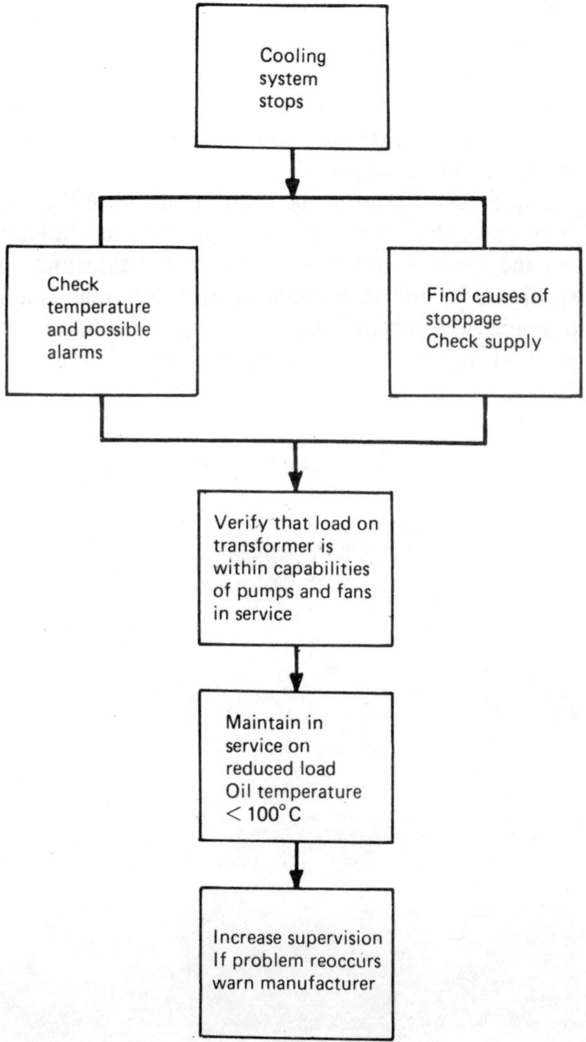

Figure 14.6. Instructions if cooling stops

transformer, the particular installation conditions, any recorded phenomena and abnormality, etc. Refer to the manufacturer when necessary for his advice on investigations to be carried out and steps to be taken.

ECONOMIC ASPECTS OF MAINTENANCE

The cost of maintenance operations, in personnel, materials and plant, increases with their complexity and frequency. This cost must be balanced against the costs which arise as a result of a transformer not being available when needed and these depend on the power and specialization of the transformer and the load it supplies. The consequences of unavailability are not the same for several transformers in parallel supplying a mechanical workshop as for one transformer connected to an electric furnace or to rectifiers in an electrolytic process.

The maintenance must therefore be carried out more thoroughly when the cost of loss of supply increases. This problem is difficult to assess precisely given the random nature of some of the information. It is the combined experience of the manufacturer and user that enables a compromise to be made between the two extremes, of making provision for a fault at any price or forgetting the transformer until it breaks down.

15 Liquid dielectrics

THE ROLE OF LIQUID DIELECTRICS

Liquid dielectrics have two essential functions: dielectric insulation and heat transfer.

Dielectric insulation

In the active part of a transformer the insulation between elements at different potentials is provided by

1. the dielectric alone when insulation is required between two bare pieces of metal – for example, switch contacts,
2. a solid layer of paper or pressboard impregnated with liquid dielectric – for example, the insulation between two conductors of the same winding,
3. mixed insulation, with layers of liquid dielectric between barriers of solid insulating material – for example, insulation between two concentric windings of different voltages.

In all these cases the contribution of the liquid dielectric is essential.

Heat transfer

Although its efficiency is exceptionally high, a transformer cannot avoid the loss of energy that accompanies all energy conversion. The energy is

Figure 15.1. Three-phase 80 MVA transformer for steel works

dissipated as heat and this leads to the need for cooling (see Chapter 5, Cooling).

Because of losses, the temperature of each element rises until an equilibrium is reached between the rate at which the heat is produced and its dissipation.

The heat lost is carried by natural or forced circulation of the liquid dielectric to the cooling equipment. A good heat transfer system can avoid the formation of hot spots by an abundant and well distributed flow, the efficiency being directly influenced by the viscosity and specific heat of the liquid dielectric used.

The two functions described above are vital for a transformer. Any defect of an electrical nature causes breakdown, the consequences of which are serious. Any abnormal heating, particularly if it is localized, results in a reduction of the electrical or mechanical properties of the solid insulating materials.

PRINCIPAL CAUSES OF CHANGES IN LIQUID DIELECTRICS

The properties of a liquid dielectric may be altered by a number of atmospheric and other common factors.

Oxygen and humidity

The importance of these has already been mentioned in the previous chapter, and their ingress into dielectrics results principally from breathing.

Oxygen has a particular influence on mineral oils and oxidation is a cause of ageing. Humidity has a harmful action on all liquid dielectrics, particularly when it is in the form of free water, in suspension, in emulsion or in combination with solid particles (dust in suspension or deposits).

Table 15.1 shows the danger presented by an accumulation of moisture in a dielectric and compares the limits of solubility of water in oil for various temperatures.

Table 15.1

Temperature (°C)	Maximum solubility of water in oil (ppm)
− 20	10
0	20
+ 20	58
+ 40	130

The moisture content is expressed as parts per million e.g. milligrams per kilo.

While a dielectric with a high moisture content can still carry out its function provided the temperature is high enough to maintain the required solubility, when sudden cooling occurs (as for example a load break in cold weather), it can become useless because of the appearance of free water in suspension or emulsion resulting from the reduction in solubility.

Temperature

Mineral oils are sensitive to accelerated ageing caused by high temperatures.

Pollution

All manner of impurities (dust, grease, etc.) can find their way into a transformer through bad handling, for example when topping up the oil.

Dust or fibres can arise from mechanical deterioration (fretting or vibration) of the solid insulating materials. Deposits sometimes occur as a result of the ageing of the liquid dielectric.

Paints, lacquers and jointing materials can sometimes be dissolved due to the strong solvent action of the dielectric, resulting in foreign matter being put into solution or in suspension. In general this problem can be avoided by a careful choice of materials.

Electrical phenomena

Alteration of the liquid dielectric because of purely electrical phenomena in transformers is virtually unknown. However, secondary effects are sometimes noticed:

Conductor chains

If dust or fibre, which may be charged with moisture, is carried by the dielectric in circulation it can, as a result of the difference of its dielectric constant from that of the liquid medium, be subjected to electrostatic attraction and be formed into a chain following the lines of force of the electrostatic field and give rise to a flashover.

Partial discharges

Under the effect of both the electric field and local variations caused, for example, by the appearance of gas bubbles due to variations of the solubility of gases in liquids, minute discharges can occur. The liquid dielectric is generally only slightly sensitive to these discharges because it is constantly replaced by circulation, but in time the discharges can become harmful if they occur on a solid impregnated insulation material and reach a level high enough to cause it to deteriorate.

MINERAL INSULATING OILS

These are obtained by distillation and refining crude petroleum. They are mixtures of many compounds known as hydrocarbons in proportions which depend on the source.

Hydrocarbons can be broadly divided into three classes, naphthenes, paraffins and aromatics. Naphthenes and paraffins are saturated hydrocarbons and are chemically stable. They differ from one another by their molecular structure and their physical and chemical properties. Aromatics are not saturated and as a result they are less stable and more chemically active.

The generally used classifications of 'naphthenic' and 'paraffinic' do not imply that these oils contain naphthene or paraffin exclusively but refer to the dominant characteristic of one of these classes in a mixture of naphthenic, paraffinic and aromatic hydrocarbons. The proportions vary according to the source of the crude and the refining process.

The sources of naphthenic crude are becoming rarer, and there will be a tendency in the future to use paraffinic crude more frequently to manufacture insulating oils. This should not cause any particular problem except from the point of view of the freezing point, and it may become necessary to use special additives, development of which is currently being carried out.

Use

Mineral insulating oils are used universally for all types of transformers of all powers and voltages, when the installation conditions, whether outside or in enclosures, enable elementary fire precautions to be taken.

Insulating oils, which were at one time subject to significant changes (oxidation and sludge deposits) due to ageing, have benefited considerably since the 1950s from improvements in processes which have given them proven stability.

Two important results of this improvement are seen in their miscibility and reduced need for inhibitors.

Miscibility: Insulating oils currently supplied by the major petroleum companies are mutually miscible in any proportion, but it is not possible to restore a used oil to a renewed condition by mixing.

Inhibited oils: Because of the high stability of present day insulating oils, the use of oxidation inhibitors to retard ageing is recommended only in exceptional cases such as in very large transformers subjected to severe operating conditions.

Characteristics

The essential characteristics of the non-inhibited insulating oils currently in use are given in the IEC standard 296 (1982). These characteristics are summarized in Table 15.6 in which the acceptable service limits set by IEC standard 422 (1973) and by our experience in operating numerous transformers of various characteristics are also shown.

OTHER DIELECTRICS

It should be noted that research is being carried out on the use of other liquid dielectric, particularly products derived from silicones.

These products will, in the future, become more widely used, but at the present time, because of their cost, they are reserved for special applications in which the results are promising. A study of this topic does not form part of this book.

CHECKING AND MAINTENANCE OF LIQUID DIELECTRICS

Check on transformers in service

This subject has already been reviewed in the previous chapter, and is only referred to again now to remind the reader that vigilance in checking the operating conditions safeguards both the dielectric and the transformer.

Heating

The temperature of the dielectric (as measured by the thermometer placed in the pocket on the cover) as a function of the variation in load and ambient temperature, must be read systematically to reveal possible abnormalities.

Transformers sometimes age abnormally because of temperature rises that occur either from repeated overloading or from poor installation conditions causing recirculation of hot air into the cooling system.

Protection of the dielectric against moisture

In the case of transformers that breathe, the load on the air dryer varies widely due to atmospheric conditions and the magnitude and frequency of variations in the transformer load. Only careful checking, carried out at frequent intervals, preferably weekly, will reveal a loss of efficiency in the air dryers and the need to carry out a regeneration.

It is not unusual to discover unexpected changes. A transformer used irregularly on low load, or even at no-load but exposed alternately to the sun and the cold may breathe more than a similar unit on constant load. A loss of load due to storm or to fog may cause saturation of the dehydrating material.

Dryers consist of a container, part of which is transparent, in which there is a quantity of dehydrating granules that act by absorption. The granules are accompanied by an indicator which is blue in the anhydrous state and pink in the saturated state. The granules can be reactivated, after saturation, by placing them in an oven for several hours at a temperature between 100 and 150°C.

It is strongly recommended that the desiccating material should be regenerated or renewed as soon as one third of it has changed colour. Airtight transformers that do not breathe do not require a dryer. However, some conservators of the membrane type do have dryers to protect them from condensation on the air side.

Checks on the dielectric during service

It has been found from experience that strict observation of the guidelines given above does ensure that the dielectric lasts over a long period. Even the best organization, however, has its faults, and it is useful at times to draw attention to them. This is why it is considered necessary to carry out periodic tests on the dielectric to follow the changes under service conditions.

It is difficult to establish an absolute rule, valid for all cases, so general guidelines are given, leaving it for each organization to adapt them to its particular needs, taking into account the possibility of interruptions to supply, the value of the installed equipment, and the cost of the tests.

It should also be noted that the results of all these tests should be retained as it is often more useful to know the changes that have occurred than to refer to limits established in absolute terms.

Frequency of testing

This frequency can normally be established on the basis of:
  2–3 years for transformers that breathe, installed outdoors or in humid buildings,
  3–5 years for transformers that breathe, installed in dry buildings,
  10 years for sealed transformers (with or without gas cushion and with a conservator of the membrane type or the flexible bag type).
A definitive list of tests cannot be compiled, but an attempt has been made to enumerate them in an order corresponding approximately to their order of priority.

The complete range of tests would be very expensive and consequently it is generally the value of the equipment and the degree of reliability required that dictates in each case the particular choice of tests.

Sampling

Whatever the test envisaged, a sample is always needed for the verification of a dielectric, and it is necessary to ensure that the sample is representative of the dielectric contained in the transformer.

As far as possible, take a sample within three hours of the transformer being taken out of use, so as to obtain an average sample resulting from the agitation due to the circulation of the dielectric. The sample will then still be warm and less likely to be contaminated by condensation of the ambient moisture. Avoid any turbulence so as to minimize contact of the dielectric with the ambient air (formation of bubbles).

Cleanliness is essential. The orifices and sampling pipework, which are often soiled with a build up of dust due to slight leakage, must be cleaned. To purge and rinse out the pipework, it is prudent to allow a quantity of dielectric, about ten times the volume of the sampling pipework, to run to waste before taking the sample.

It is advisable to fill the sample container by means of an immersed tube and allow it to overflow by an amount equal to the volume of the container to rinse it out.

Finally, care must be taken that the material of the sample container and any other apparatus used during sampling does not react with the dielectric. From this point of view, glass is the best material.

The IEC standard 475 (1974) 'Methods of Sampling Liquid Dielectrics' should be consulted.

Electric strength (breakdown voltage)

This is the most commonly known test applicable to mineral insulating oils, and it was originally developed for this purpose. It should always be included.

Method of measurement   This is standardized by IEC 156 to which one should refer. An oil test cell is used in which an alternating voltage is applied between two metal spheres 12.5 mm in diameter with a gap of 2.5 mm between them. The voltage is increased until breakdown occurs.

The flashover must be quickly stopped to allow six successive measurements of the rupture voltage on the same sample.

The value of the electric strength of the sample tested is the average of the six measurements. According to the French standard NF C 27 221 it is the average of the last five of the six measurements, but if there is a difference greater than 20 per cent between one of the measurements and the average then it is considered prudent to carry out a further measurement to verify that the difference was not due to an error in carrying out the test.

Note: The standard requires the oil in between the electrodes to be lightly stirred between two measurements, using a clean, dry glass rod. Some test cells use a small pump which is difficult to clean and sometimes causes errors due to mixing of liquids from two successive tests. We prefer to use an immersed agitator of the magnetic drive type, which introduces into the liquid to be tested only a simple moving part which is easy to clean.

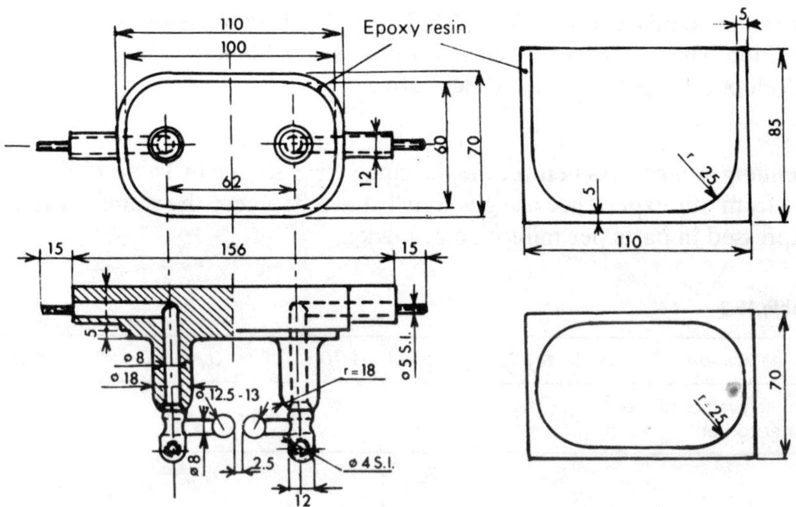

Figure 15.2. Oil test cell

Limiting values   The permissible limits of the dielectric strength of the dielectrics used in transformers under service conditions are shown in Table 15.6. These are shown in Table 15.2 to draw attention to the fact that they differ sometimes from reference documents but they correspond with our experience in the operation of numerous transformers.

Treatment   As a general rule, a drop in the dielectric strength is due to pollution of the dielectric by dissolved moisture and/or dust in suspension.

The various treatment processes, which are used according to the degree of pollution and the quality level to be re-established, aim to carry out simultaneous filtering and de-humidifying of the dielectric liquid (see later).

It is desirable, by means of a preliminary treatment during the first filling of a transformer, to improve the dielectric strength of the dielectric to voltage levels 10 kV above the minimum voltages in Table 15.2.

Table 15.2

Rated Voltage	$U_n \leqslant 36\,kV$	$36 < U_n \leqslant 70$	$70 < U_n \leqslant 170$	$170 < U_n$
Minimum values of the dielectric strength for oils	30	35	40	50

Water content

Method of measurement   As only very small quantities of water are involved (5–100 ppm), the measurement involves the Karl Fischer method, as in the standard ISO 12760 (NF T 60 113). It consists of an iodometric titration which can only be carried out in special laboratories. It has been developed for petroleum and derivatives.

Limiting values   As before, the limiting values shown in Table 15.6 resulting from our experience are given in Table 5.3, where the water content is expressed in parts per million, e.g. mg/kg.

Table 15.3

Rated Voltage	$U_n \leqslant 36\,kV$	$36 < U_n \leqslant 70$	$70 < U_n \leqslant 170$	$170 < U_n$
Limiting values of the water content for oils	40	35	30	20

Treatment   The moisture present in a dielectric does not necessarily have an immediate and direct effect on the dielectric strength, particularly if the concentrations remain below saturation, but it does create a risk.

The removal of this moisture requires rather complicated processes (vacuum treatment) particularly when it is required to reduce the moisture content below 20 ppm.

It is desirable during the first filling of a new transformer to reduce the moisture content to 10 ppm below the values given in Table 15.3.

Dielectric loss factor (tan δ)

Method of measurement   This method is standardized in IEC 247. It employs a measuring cell forming a capacitor in which a 50 Hz alternating current is used to evaluate the leakage current I_r and the capacitive current I_c from which the ratio $I_r/I_c = \tan \delta$ gives the loss angle.

Limiting values   As the value of the loss factor varies with temperature, it is recommended that the test be carried out at the reference temperature of 90°C. If, however, it is necessary to carry it out at 25°C the reference values for the oils must be divided by 15. This is only approximate since the variation of the loss factor with the temperature is dependent on several parameters which do not all act in the same way when the temperature varies.

As before, the limiting values shown in Table 15.6 resulting from our experience are given in Table 15.4.

Treatment   The increase in the dielectric loss factor has several causes. Often filtration and dehydration have a beneficial effect on the loss factor.

Table 15.4

Rated Voltage	$U_n \leqslant 36\,kV$	$36 < U_n \leqslant 70$	$70 < U_n \leqslant 170$	$170 < U_n$
Limiting values of tan δ for oil at 90°C and 50 Hz	1.5	0.8	0.3	0.2

This can be the case particularly during the first filling of a new transformer.

However, there are occasions when dielectrics, heavily polluted as a result of ageing or an internal fault, cannot be restored within acceptable limits of the loss factor without recourse to reconditioning by physical-chemical means similar to expensive refining processes (see later).

Neutralization value (or acid index)

Method   This method is specified in IEC 296 for insulating oils. The value is expressed as milligrams of potassium necessary to neutralize the total acidity in one gram of sample.

Limiting values   The acidity of dielectrics, at the low contents normally met, has no direct effect on the behaviour of the dielectric but is a warning sign of degradation.

The neutralization value for insulating oils increases progressively with ageing. It is not correct therefore to refer to limiting values for the neutralization value, but a safety limit of 0.5 mg KOH/g, above which unexpected alteration is possible.

Treatment   When the neutralization value increases to the point where the preservation of the dielectric is at risk, it is prudent to replace it rather than proceed with an expensive treatment. (Sometimes the supplier of the dielectric can carry out regeneration.)

Interfacial tension

Method   This method is specified in ISO 6295. It applies only to insulating oils.

The measurement consists of evaluating (in mN/m) the force necessary to pull an oil/water interface through a metallic ring in predetermined conditions. This force, which is related to the phenomenon of capillarity, varies with the composition of the oil and is affected by the products of decomposition.

Limiting values   The interfacial tension of an oil varies with its degree of ageing and, as with the neutralization value, it can be considered as a warning sign that changes are occurring.

The values in Table 15.6 shown again in Table 15.5 are the empirical lower limits.

Table 15.5

Rated Voltage	$U_n \leq 36\,kV$	$36 < U_n \leq 70$	$70 < U_n \leq 170$	$170 < U_n$
Minimum value of the interfacial tension of oil	10	12	15	20

Treatment   An abnormal drop in the interfacial tension of an insulating oil is a characteristic sign of ageing and it signals quite important physical and chemical changes. In this case it is preferable, as in the preceding case, to replace the oil rather than treat it.

Analysis of gases dissolved in oil

This measurement, which has only relatively recently been available, consists of chromatographic analysis of the dissolved gases in a sample of oil. It detects at an early stage phenomena such as local or general heating, partial discharges, etc., which are likely to affect the life of a transformer. It is outside the range of this book, but it is mentioned because of the service it can render particularly for large transformers at high voltage.

REPLACEMENT, REGENERATION OR TREATMENT OF LIQUID DIELECTRICS

Replacement of the dielectric

Replacement of the dielectric may be required when it is rejected or just polluted to the point where any treatment is more expensive than replacement.

When the dielectric is replaced, the active part of the transformer must be thoroughly rinsed to remove sludge or deposits, with the assistance of suitable solvents or a small quantity of new dielectric, which is then thrown away.

The rejected products must not be disposed of onto the ground or into a sewer because of risks of pollution. Mineral oils of a composition similar to a fuel can be disposed of by combustion or they can be returned to waste recovery treatment specialists.

Replacement of the dielectric is rarely absolutely necessary. One instance of when it must be done is when the oil is excessively aged, particularly when the neutralization value is above 1 mg KOH/g and the interfacial tension is below 10 mN/m.

Regeneration

Regeneration is necessary when a dielectric is quite severely affected by pollution from decomposition products. If these are insoluble, they cause deposits which can only be removed by emptying the tank, and if they are soluble they contaminate the dielectric and sometimes act as catalysts in further decomposition. They can only be removed by chemical or physical and chemical methods similar to refining.

The costs of arranging and carrying out this regeneration are quite out of proportion to the value of the dielectric, and it is for this reason that, as a general rule, it is better to use specialist firms or the supplier of the dielectric.

Treatment of the dielectric

Fortunately, in the majority of cases, a polluted dielectric is contaminated by moisture, suspended matter or dissolved gas and removal is possible by using one of the many methods described below.

Screens or sieves

These are constructed of metal, fabric or textiles, mesh formed of thin metallic sheeting or spirally wound wire. They retain solid particles of dimensions larger than those of the mesh, but they have only low efficiency with the fine particles which are the most harmful for transformers due to the ease with which they remain in suspension.

These screens are still used, however, as prefilters or strainers to delay the clogging of the filters themselves.

Filter presses

These consist of a series of sheets of porous paper or felt on a perforated metal support arranged in such a way that there are groups of sheets in series (multiple filtration) or in parallel (large area) in which the fluid to be

Table 15.6 Insulating oils

Characteristics		New oils		Acceptable limits for oils during service			
		NF C27 101-IEC 296		NF C 27 222 – IEC 422			
		Class	*Class*	*For rated voltages*			
		I	*II* ▲	$U_n \leqslant 36\,kV$	$36 < U_n \leqslant 70$	$70 < U_n \leqslant 170$	$170 < U_n$
Density at 20°C		$\leqslant$0.895	$\leqslant$0.895	—	—	—	—
Kinematic viscosity at	40°C	$\leqslant$16.5	$\leqslant$11.0	—	—	—	—
(centistokes)	−15°C	$\leqslant$800	—	—	—	—	—
	−30°C	—	$\leqslant$1800	—	—	—	—
Pour point (°C)		$\leqslant$−30	$\leqslant$−45	—	—	—	—
Flash point	Closed vessel (°C)	$\geqslant$140	$\geqslant$130	$\geqslant$115	$\geqslant$115	$\geqslant$115	$\geqslant$115
Neutralization index (mg KOH/g)		$\leqslant$0.03	$\leqslant$0.03	$\leqslant$0.5*	$\leqslant$0.5	$\leqslant$0.5	$\leqslant$0.5
Moisture content (ppm)		●	●	$\leqslant$40*	$\leqslant$35*	$\leqslant$30	$\leqslant$20

Breakdown voltage (kV)	on oil sampled	$\geq$30	$\geq$30	$\geq$30*	$\geq$35*	$\geq$40	$\geq$50
	on oil filtered and dried	$\geq$70*	$\geq$70*	—	—	—	—
Dissipation factor at 50 Hz and 90°C	on oil sampled	—	—	$\leq$1.5*	$\leq$0.8*	$\leq$0.3*	$\leq$0.2*
	on oil filtered and dried	$\leq$0.005	$\leq$0.005	—	—	—	—
Interfacial tension at 25°C (mN/m)		$\geq$40*	$\geq$40*	$\geq$10*	$\geq$12*	$\geq$15*	$\geq$20*
Characteristics after oxidation (NF C27 220-IEC74)	Sludge %	$\leq$0.10	$\leq$0.10	Average calorific value between 25 and 125°C: 0.5 cal/g/°C			
	Neutralization value (mg KOH/g)	$\leq$0.40	$\leq$0.40	Average coefficient of expansion between 25 and 125°C: 7.10^{-4} per °C			

Notes: ▲ In general, insulating oils class II are preferred for cold climates.
     ● Before filling a new transformer, a new oil must be treated so that its moisture content is 10 ppm less than that allowed in service.
     * We have interpreted or modified certain values shown in the standards in accordance with our experience.

treated passes through the filter medium perpendicular to its surface. The filtration characteristics depend therefore on the porosity of the paper or the felt used.

Filter presses remove suspended matter quite effectively and to a certain extent they dry the dielectric by retaining the free water by capillarity and absorbing the dissolved water (so long as the filter medium is itself dry). However, they become saturated quite rapidly and they cannot operate continuously as they must be periodically dismantled to have the filter elements replaced by new ones or at least be cleaned and dried.

Filter cartridges

These cartridges usually consist of a stack of discs (paper or fabric) through which the fluid passes from the outside to the centre, parallel to the surface of the material. The filtration characteristics therefore depend as much on the material used as on the pressure exerted on the stack of discs.

The material of the cartridges can also remove the water, but the effect is often limited because of the high pressures used to circulate the dielectric.

Filter beds

Associated with the filters described above, there may be used finely divided material such as activated alumina, clays or Fullers' earth, having a selective physical-chemical action of absorption of the dissolved water and certain chemical impurities, thus improving the purification.

However, they are used only in large installations.

Centrifuges

Separation of solids from liquids, or of a mixture of liquids of different densities, can only be carried out slowly by settlement.

Centrifuges subject mixtures to high accelerations (several hundred times g) by rotation in a bowl, and the separation is carried out in a very short time. The arrangement of the equipment (multiple conical bowls) enables them to operate continuously at least for liquid/liquid separation (water and oil).

As the solid impurities remain at the bottom of the bowl, prefiltration is necessary to avoid blockage.

Some centrifuges capable of operating with hot oil and under vacuum can satisfactorily remove some of the dissolved moisture.

Dewatering, degassing columns

By subjecting a hot finely dispersed liquid in thin layers (columns of plates or Raschig rings) to a sufficiently low vacuum (limited by the boiling point of the liquid), the water and most of the dissolved gases can be removed.

This process is frequently used for oil having the vapour pressure considerably lower than for water.

Combined installation

The processes described above are rarely used alone but usually in combination. The most usual are shown below in approximate order of their efficiency.

Prefilter + filter press.
Prefilter + centrifuge.
Prefilter + vacuum centrifuge.
Prefilter + cartridge filter + dewatering and degassing column.

INTERNATIONAL STANDARDS

IEC 475 (1974): Methods of sampling liquid dielectrics.
IEC 247 (1978): Measurement of relative permittivity dielectric dissipation factor and DC resistivity of insulating liquids.
ISO 760 (1978): Determination of water by Karl Fischer method (general method).
ISO 6295 (1983): Determination of interfacial tension using oil against water-ring method.
IEC 296 (1982): Specification for unused mineral insulating oils for transformers and switchgear.
IEC 422 (1973): Maintenance and supervision guide for insulating oils in service.
IEC 74 (1963–73–74): Methods for assessing the oxidation stability of insulating oils.
IEC 474 (1974): Test methods for oxidation stability of inhibited mineral insulating oils.
IEC 156 (1963): Methods for determination of the electric strength of insulating oils.

IEC 733 (1982): Determination of water in insulating oils and in oil-impregnated paper and press board.
IEC 567 (1977): Guide for the sampling of gases and of oil from oil filled electrical equipment and for the analysis of free and dissolved gases.

16 Over-currents

INTRODUCTION

Chapter 6 explained overloads and the way in which transformers could be designed to withstand them. This chapter deals with over-currents, which are distinguished from overloads by their short duration (from tenths of a second to tens of seconds) and by their relative intensity (2–25 times the rated current).

Over-currents are principally of the following nature:

1. short circuit at the secondary $I_{cc} = 6$ to $25\ I_n$,
2. motor starting currents $I_d = 3$ to $7\ I_n$,
3. switching-in of transformers $I_o = 3\sqrt{2}$ to $8\sqrt{2}\ I_n$.

The effects of these surges fall into three categories:

1. electrodynamic forces involving high stress in the transformer windings,
2. high temperature rise of the winding,
3. drop in secondary voltage.

The behaviour of the transformer in these three categories will be examined. Some simple formulae will help to understand the influence of the various parameters.

SHORT-CIRCUIT IMPEDANCE OF A TRANSFORMER

The short-circuit impedance E_{cc} of a transformer is used in the calculation of the short-circuit current and of the voltage drop during an over-current.

Figure 16.1. 29 MVA, 220 kV transformers to supply the auxiliaries of a 900 MW generator unit at a nuclear power station.

E_{cc} can be considered as the resultant of vectors in quadrature

$$E_{cc} = \sqrt{(RI\%)^2 + (XI\%)^2}$$

where $RI\%$ is the resistive component of E_{cc}, usually negligible compared
with $XI\%$, and

$XI\%$ is the reactive component of E_{cc}.

$$XI\% = K\left(\varepsilon + \frac{a_1 + a_2}{3}\right) \times \frac{D_m}{H}$$

The distance between the HV and LV windings has a value imposed by the
strength of the insulation subjected to the specified dielectric stress (BIL).
The economic optimization of the transformer on the other hand is related
to the ratio H/D_m while the radial thicknesses a_1 and a_2 are proportional to
the rated power.

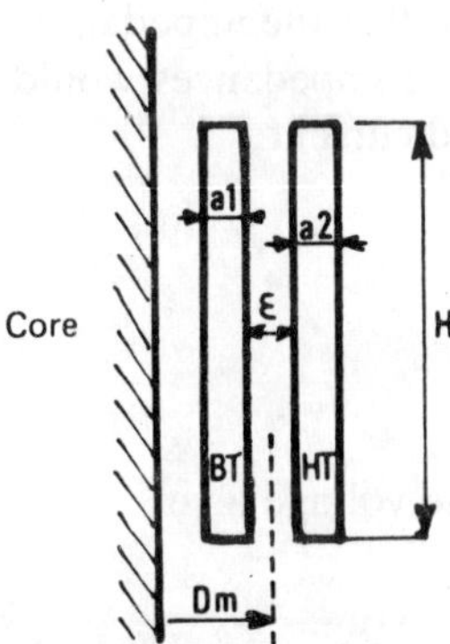

Figure 16.2.

The term $XI\%$ thus varies within very close limits in the case of normal
transformers. These variations are given (in percentages) in Table 16.1.

It is always possible to construct a transformer departing from these
natural values of $XI\%$, but it will be less economic. The increase in cost
brought about by specifying a short-circuit impedance differing from the
natural value must be compared with the technical and economic advantages which result.

Table 16.1

P (MVA)	U (kV)	7.2	24	72.5	170	245	420
0.1–1		4%	5%				
1–10			7%	9%			
10–100				10%	12%	13%	
100–1000					12%	13%	15%

CALCULATION OF SHORT-CIRCUIT CURRENT

Symmetrical component

The impedance in ohms per phase of a transformer in three-phase short circuit at the secondary terminals is equal to

$$Z_t = \frac{U^2}{P} \times \frac{E_{cc}\%}{100}$$

where U is the line voltage of the system in kV,
P is the rated power in MVA, and
E_{cc} is the short circuit impedance in percent related to the rated power.

To calculate the current in a transformer on short circuit, the impedance per phase of the supply must be taken into account, $Z_R = U^2/P_{cc}$ (where P_{cc} is the short-circuit power of the system in MVA) as well as the impedances of the connections (lines and cables). To ignore these impedances would lead to an increase of 10–20 per cent in the estimated current.

$$\text{Finally, } I_{cc} = \frac{U}{\sqrt{3} \times \sum Z}$$

$$\text{where } \sum Z = \frac{U^2}{P \times 100} \left(E_{cc}\% + 100\frac{P}{P_{cc}} \right)$$

Note: all impedances must be calculated for the same voltage level.

Calculation for the peak value of the short-circuit current

Depending on whether the short circuit occurs as the system voltage passes its maximum or zero, the current will be symmetrical or asymmetrical. The peak value of the current will therefore be very different in the two cases (Figures 16.3 and 16.4).

CALCULATION OF THE MECHANICAL FORCES AND STRESSES RESULTING FROM SHORT CIRCUITS

Laplace's law states that if an element dl of a conductor in which a current i flows is in a field of flux density B, the force dF on the element of the conductor is given by

$$\vec{dF} = i\,\vec{dl} \wedge \vec{B} \text{ (see Figure 16.5)}$$

The calculation of the forces in a winding requires a knowledge of the field B at all points in the winding. This knowledge is only available by manual

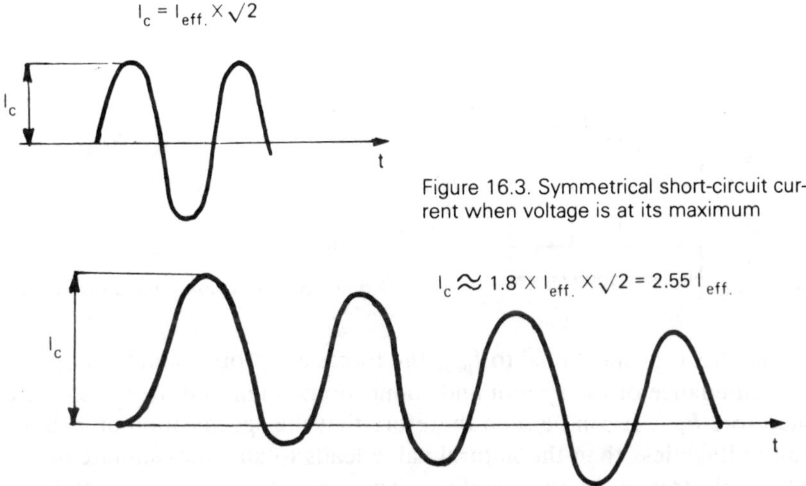

$$I_c = I_{eff.} \times \sqrt{2}$$

Figure 16.3. Symmetrical short-circuit current when voltage is at its maximum

$$I_c \approx 1.8 \times I_{eff.} \times \sqrt{2} = 2.55\, I_{eff.}$$

Figure 16.4. Fully asymmetrical current when voltage is near to zero

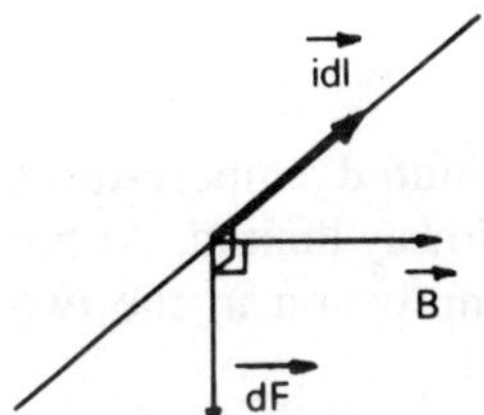

Figure 16.5. Three dimensional representation of vectors $i\,dl$, B, dF

calculation at the axial mid-height of uniform windings where the vector B is axial and the forces purely radial.

At the extremities of the winding, the lines of force B curve and it is necessary to find the components B_{ax} and B_{rad} to calculate respectively F_{rad} and F_{ax}. These calculations can be carried out only with the aid of a computer. Nevertheless some examples will help to understand the direction of these forces.

Radial forces

The flux at mid-height of the windings is purely vertical (see Figure 16.6). Its value is maximum near to the gap separating the HV and LV windings.

It decreases linearly to zero on the outside of the exterior winding and the inside of the interior winding.

The forces arising from this axial field of flux are purely radial and for the two windings they are mutually repellent: the exterior winding is subject to tensile forces while the interior winding is subject to compressive forces.

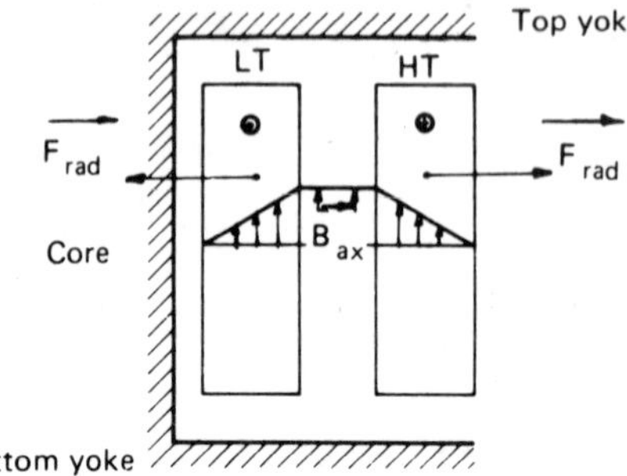

Figure 16.6. Axial flux in uniform windings

As the flux is proportional to I_{peak} the forces are proportional to $(I_{\text{peak}})^2$. If the impedance of the system and connections is ignored the stresses are related to $1/E_{cc}^2$. It can be seen therefore that the specification of a short-circuit voltage less than the natural value leads to an over estimate of the mechanical stresses in the windings when a short-circuit current flows though them.

Axial forces

For windings of equal height and uniformly distributed ampere-turns (Figure 16.7), the length of the actual winding is obviously limited. At the ends the lines of force B curve and cause axial compression in the two windings.

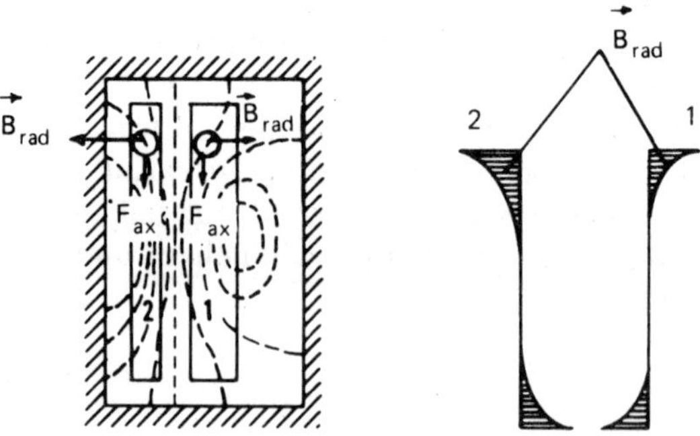

Figure 16.7. Windings of equal height and uniformly distributed ampere-turns

For windings of unequal height and displaced centres (Figure 16.8), it will be necessary to add to the forces referred to above those caused by possible displacement of the centres in the axial direction. Winding No 1 is pushed down and No 2 is pushed up. These two forces are equal in strength.

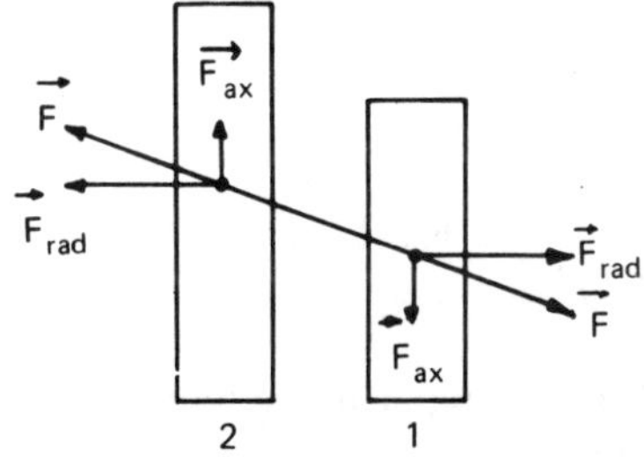

Figure 16.8. Forces on supports in the case of windings of unequal height and displaced centres

Layouts improving the behaviour in short circuit

Radial forces

Exterior winding   The stresses must not cause inadmissible extension of the conductors, because of the risk of short circuit with the tap winding usually placed outside the main winding, or even between neighbouring phases.

The winding operation must be carried out so as to ensure maximum tightness and thus avoid the risk of short circuit between turns due to radial vibrations.

Interior winding   The compressive stress may lead to buckling of the conductors (resistance of a ring subjected to external force). Solid radial supports are required. They often take the form of a rigid insulating cylinder. The resistance to compression is sometimes increased by impregnation of the conductors.

Axial forces

Internal compressive forces   The values of these forces can exceed the initial clamping pressure. This is why the windings are often subjected to compressive vibrations. The spacers and spacing segments must therefore be radialized to avoid deterioration of the layers of paper covering the conductors. The oil impregnating the spacers has the effect of acting as a damper and limits the amplitude of the vibrations.

Forces on the supports   The layout of the winding must be such that, whatever the position of the tappings, the forces on the supports do not exceed the allowable limit.

In addition, the initial clamping pressure must exceed these forces. Under these conditions the winding remains unaffected by the oscillatory forces and there is no risk of amplification.

Stabilization of the windings must be carried out in the workshop before final wedging in such a way that displacement of centres between the windings is restricted to very fine limits in order to minimize the forces on the supports.

CALCULATION OF TEMPERATURE RISE DUE TO OVER-CURRENTS

Suppose the windings of a transformer are at an initial temperature of θ_i and that they are subjected to an over-current for time t.

This time can be regarded as being sufficiently short for the heating effect in the windings not to be dissipated. These losses will therefore cause an adiabatic temperature rise in the conductors.

The final temperature θ_f is given by:

$$\theta_f = \theta_i + a\,J^2\,t\,10^{-3}$$

where $a \approx 8$ for copper windings,

$\quad t = $ time in seconds,

$\quad J = $ current density in $A_{\text{eff}}/\text{mm}^2$ during the over-current period.

If n is the coefficient of overload ($n = I/I_n$), J can be taken as $4n$ as a first approximation.

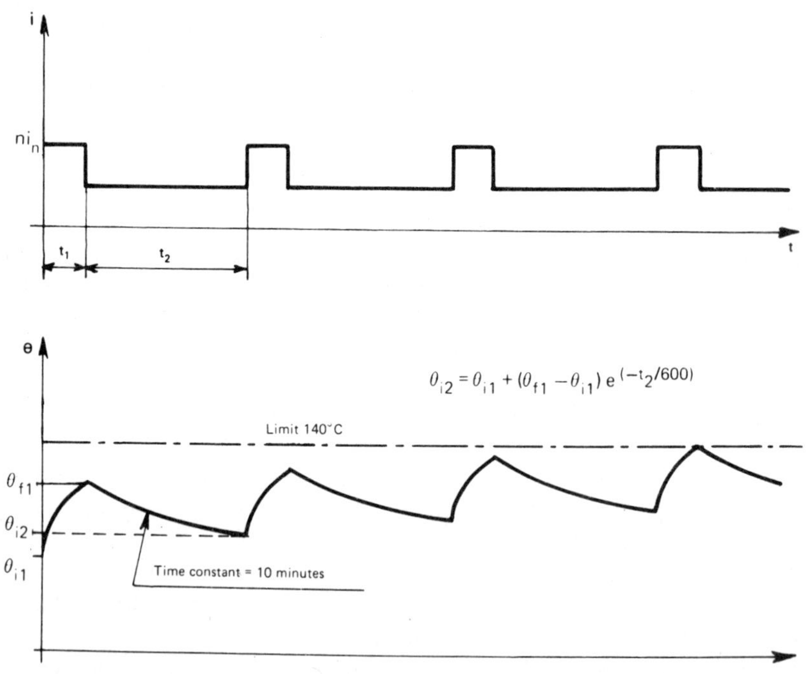

Figure 16.9.

The formula thus becomes

$$\theta_f = \theta_i + 0.128\, n^2 t.$$

As soon as the surge ends, the conductors lose the accumulated heat across the insulation to the liquid dielectric. The thermal time constant for the windings is about ten minutes. Therefore half the temperature rise $\theta_f - \theta_i$ calculated above will be dissipated in ten minutes.

The temperature rise after 3 seconds of over-current must not exceed 250°C. For longer durations or more frequent surges, this limit must be reduced to 140°C.

From the initial temperature θ, the coefficient of overload n, the duration of over-currents t_i, with time intervals t_2, the number of cycles that must not be exceeded can be calculated (Figure 16.9).

Premature ageing of the insulation and an excessive shortening of the life occur if the limit of 140°C is exceeded too frequently.

VOLTAGE DROP DUE TO OVER-CURRENTS

The voltage drop due to an over-current of factor n is given by

$$\Delta U\% = n(RI\%\, \cos \phi + XI \sin \phi)$$

where $RI\%$ is the resistive component and $XI\%$ is the reactive component of E_{cc}. ϕ is the phase angle between current and voltage.

At a constant $\cos \phi$ the voltage drop is therefore proportional to the over-current factor n.

This voltage drop is considerable when direct starting an asynchronous motor.

Advantages of a high short-circuit impedance

The secondary voltage at the terminals of a transformer is equal to $V\psi$

$$U\left(1 - \frac{\Delta U\%}{U}\right).$$

This voltage drop reduces the starting current I_d of the motor which becomes equal to $I = I_d\left(1 - \dfrac{\Delta U}{100}\right).$

If the load torque during the starting period is not too high (as with fans and centrifugal pumps), the starting time t_d will be increased in such a way that the product $I^2 t_d$ is virtually constant.

In this case $t_d = t_{d\,\text{nominal}}\left(1 + 2\,\dfrac{\Delta U\%}{U}\right).$

Protection by limiting $I_{\max}$ is thus not so necessary at the starting peaks, and the selection of protection is easier to make.

The natural reduction of I_d due to a high E_{cc} makes it possible to do without progressive star-delta starting systems, stator or rotor resistances, auto-transformers, etc.

The no-load voltage of the transformer should be specified so that, on subtracting the voltage drop due to the normal load, the rated voltage of the motor is obtained.

Warning: this starting system can only be used if the load torque is maintained at normal speed (load removal risks continuous over-voltage causing high temperature rise in the magnetic circuit and, by thermal conduction, in the insulation of the stator windings).

Disadvantages of voltage drop

Two disadvantages of voltage drop are that when the resisting torque is independent of the speed (e.g. in a crusher) the reduction of starting torque may cause the motor to stall, and an increase in the starting time sometimes leads to unsatisfactory operation (e.g. a motor for a cooling pump in a nuclear power station).

TRANSIENT CURRENTS DUE TO SWITCHING IN A TRANSFORMER

Sudden application of voltage to a transformer causes a unidirectional current I_c to pass whose magnitude depends on the phase angle at the moment of switching and the remanent flux density in the magnetic circuit (polarity and magnitude).

In the most unfavourable case, I_e can reach from $3\sqrt{2}$ to $8\sqrt{2}\ I_n$ as a peak value. This current will decrease to the rated current I_n within a maximum of one second.

This current circulates in one phase. It reaches half this value and flows in the opposite direction in the two other phases.

It is very much higher when the voltage is applied to the winding nearest the core.

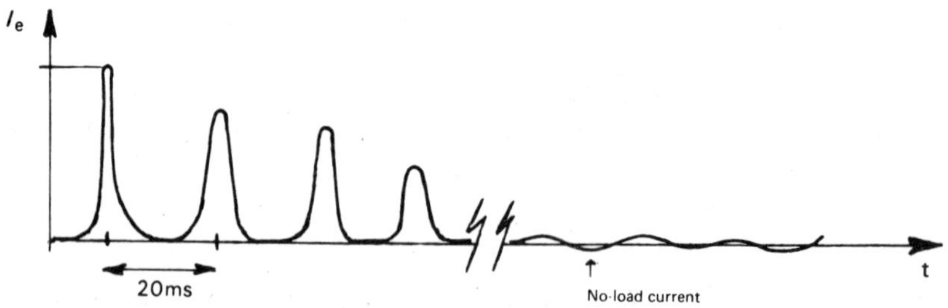

Figure 16.10.

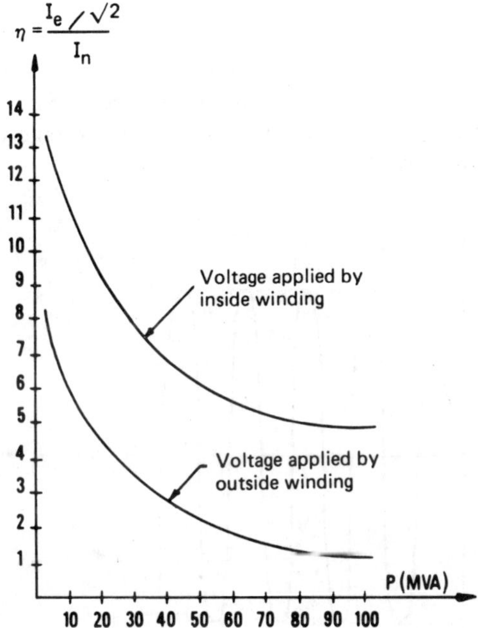

Figure 16.11. Values of switching-in currents as a function of rated transformer power for grain-oriented steel

Effects due to switching-in currents

Mechanical stresses

The winding in which the switching-in current flows tends to be compressed vertically and to expand radially (see Figure 16.12). There is no force on the support. The stresses rarely exceed those in short circuit.

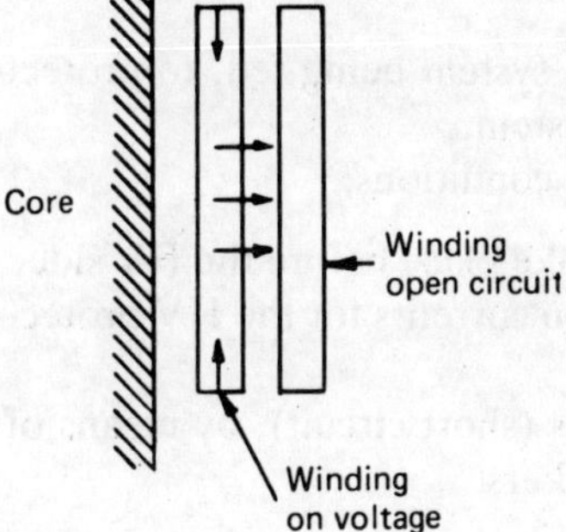

Figure 16.12.

Voltage drop

The secondary voltage of a transformer in star–star drops in the phase subjected to the greatest over-current and on one half of the voltage wave (see Figure 16.13). In the other two phases the voltage drop is half and occurs at the same time.

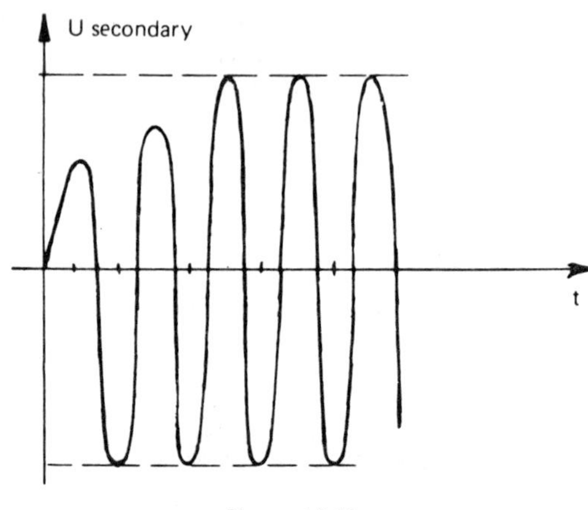

Figure 16.13.

Note: if a motor is permanently electrically connected to the transformer secondary, the voltage drop referred to above is increased by that due to the starting current of the motor.

An advantage is the longer starting time. Disadvantages are the delicate control necessary of the primary protection (accumulation of transient phenomena), and the risk of stalling the motor.

PROTECTION AGAINST OVER-CURRENTS

The protection against over-currents was discussed in Chapter 8, Protection.

Protection has a triple role: to protect the system being fed, to protect the transformer, and to protect the supply system.

Protection must comply with the following conditions:

be selective: tripping of the LV side must take place before the HV side,
allow normal overloads to pass: switching-in currents for the HV protection, starting currents for LV protection,
be rapid and if possible restrict high surges (short circuit), by means of HV fuses and LV fuses and/or circuit-breakers.

17 Over-voltages

From the consideration of over-currents in the previous chapter, we move on to the important problem of over-voltages affecting transformers.

The study of over-voltages falls within the general aim of insulation coordination, which is the subject of IEC publications 71–1 and 71–2.

Figure 17.1. 25 MVA transformers installed in a 90/21 kV substation

Over-voltages are examined under the following headings: generation, parameters affecting their magnitude, protective devices, and the influence of the insulation level on the price of transformers.

INSULATION COORDINATION

IEC 71–1 defines insulation coordination as follows:

> 'Insulation coordination encompasses the selection and application of a dielectric strength for all parts of the electrical equipment, taking into consideration the voltages of the system for which the equipment is intended and the type of equipment available. Its aim is to reduce to a minimum level, acceptable from the point of view of economy and operation, the probability that the resulting dielectric stresses on the equipment will cause damage to the insulation or affect continuity of supply.'

In service, electrical equipment can be subjected to stresses falling into the following classes:

lightning impulses: wavefront between 0.1 and 10 μs,
switching over-voltages: wavefront between 100 and 500 μs,
temporary over-voltages, at power frequencies: wavefronts longer than 500 μs.

(It should be remembered that a quarter of the fundamental wave at 50 Hz represents a wave of 5000 μs.)

These three classes will be considered in turn.

ATMOSPHERIC OVER-VOLTAGES

Origin and formation

These over-voltages originate from atmospheric discharges which act directly by striking one or more line conductors, or indirectly by striking near to the line, the ground or metallic structures, sometimes by induction, sometimes by return strokes.

The level of these over-voltages, which stress the insulation of the substation and particularly that of the transformers, depends on the line construction and the configuration of the system.

The substation circuit is a major factor if the transit time of the over-voltage to the interior of the substation is not negligible compared with the duration of the over-voltage wave (see the next section).

For systems of rated voltage greater than 52 kV, lightning strokes which cause appreciable over-voltages are either direct strokes on the phase conductors or strokes on pylons or earth wire followed by a return stroke.

For systems with rated voltage below 52 kV, attention must also be paid to induced lightning strikes. Also within this range of voltages must be taken into acount over-voltages transmitted through transformers from a high voltage system.

The conventional wave universally accepted as representative of atmospheric over-voltages, is the standard full wave called 1.2/50. The flashover of a spark gap or of an insulator chain changes this wave into a 'chopped wave' of which the most representative is one chopped after 3 μs.

Parameters influencing the level of impact on a transformer

Over-voltage waves of atmospheric origin which appear at a point in the system are impulses (full wave) or a succession of two impulses of opposite direction (chopped wave) and their transmission obeys those laws relating to high frequency transmission lines (travelling waves).

The electrical properties and the length of an overhead line or underground cable, assumed to be loss free, can be defined by three characteristics:

velocity of propagation V (V_L = 300 m/μs for an overhead line and V_C = 150 m/μs for a cable),
the surge impedance Z (Z_L = 500 Ω for an overhead line and 10–15 times less for a cable; Z_C = 30–50 Ω),
the time of propagation T which can be deduced from the length L: $T = L/V$.

It is also known that when an impulse travels along a uniform line it suffers little change, other than some attenuation due to losses, provided there is no variation of impedance.

It is sometimes said that an over-voltage originating along the line and which has been limited by flashover to earth or across an insulator becomes attenuated in amplitude and steepness of wavefront on arrival at the substation or the transformer.

Unfortunately there is sometimes an abrupt change in the type of pylon, or a change from overhead line to underground cable, at the entry to a substation which causes a change in the surge impedance. This is the cause of reflection phenomena, which are sometimes evident at the input of the transformer itself (high input impedance to impulse voltages) when the transit time of the last protected section can no longer be considered negligible compared with the time of the wavefront.

This consideration makes it necessary to place the protective devices, arresters for example, as close as possible to the transformer input terminals, and whatever the circumstances no further away than 20–30 metres for overhead lines or 10–15 metres for underground cables. (The steepness of chop for the chopped waves, which is of the order of about 0.2 μs wavefront, must be taken into account, and therefore restricts the time of propagation in the section considered to a value of less than 0.1 μs.

It sometimes happens that the line cannot act in a limiting role, as when lightning strikes the three conductors of a line mounted on wooden poles. Three travelling waves arrive simultaneously at the input to the substation at a level higher than normal for earthed metal supports. Even if each of the three over-voltages is suitably restricted, the most serious consequence of this particular case of three pole impulse occurs when they pass simultaneously through the windings of all three phases of the transformer. When they arrive at the neutral point, if the latter is not earthed through an impedance which is sufficiently low to high frequencies, they cause dangerous oscillations by reflection which can double the initial over-voltage.

This justifies the addition of an arrester at the neutral point to the standard line arresters.

TRANSIENT OVER-VOLTAGES DUE TO SWITCHING

Switching out

Transient over-voltages can arise when an inductive load is switched out of circuit, such as transformers on no-load, reactors, or reactors supplied by a transformer even when the current that is cut does not exceed 100 A.

A transformer on no-load is represented by its magnetizing reactance

$$L = \frac{U}{100\pi i_0} \quad \text{(if the frequency is 50 Hz)}$$

where i_0 is the no-load current, which for practical purposes equals i_L as at the rated voltages the current i_c is neglibible compared with i_L.

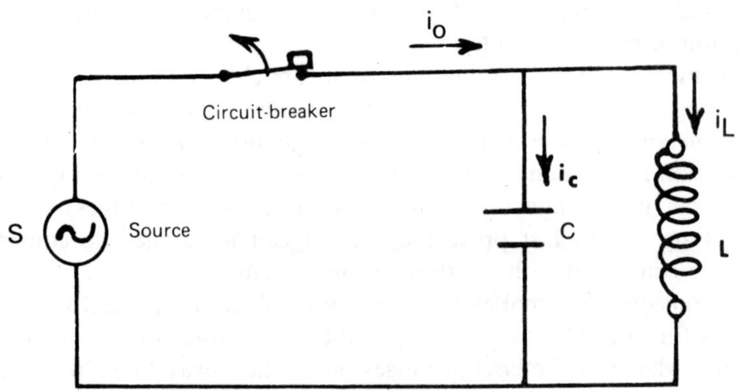

Figure 17.2

The capacitance C in Figure 17.2 is the equivalent capacitance of the transformer and it results from the combination of the series capacitances C_s, between turns, and the shunt capacitances C_d, between winding and earth, of the winding supplied (the capacitance of that part of the line between the circuit-breaker and winding must also be included in C).

When its contacts open, the circuit-breaker risks prematurely chopping the current flowing through it before it attains its natural zero value.

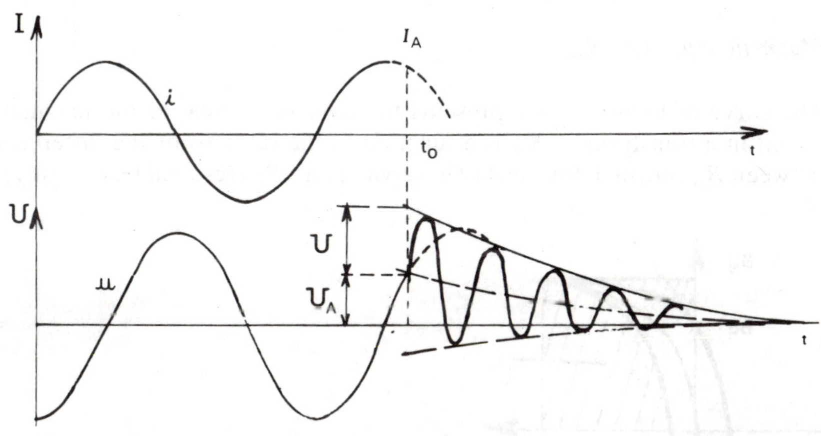

Figure 17.3.

Before the circuit-breaker is opened, the source S maintains the circuit in a state of forced oscillations. Between S and LC there is an oscillating exchange of energy at a frequency double that of S.

At the instant T_0, the circuit-breaker chops the current I_A, the exchange of energy is abruptly stopped, and the residual energy shows in the form of an exchange between L and C which rises to an oscillating voltage U with frequency $1/(2\pi\sqrt{LC})$ (self resonant frequency of the circuit $= f_p$).

The energy exchanges between inductance and capacitance can be equated:

$$\tfrac{1}{2}LI_A^2 = \tfrac{1}{2}CU^2$$

from which $U = I_A\sqrt{(L/C)}$.

In practice, a factor K_m is introduced which is called the magnetic efficiency and is dependent on the nature of the inductance.

$$U = K_m I_A \sqrt{(L/C)}$$

$K_m = 0.2$–0.4 for a transformer and 1 for a pure reactor.

It can also be shown that the maximum over-voltage is obtained when the circuit-breaker chops the current near its maximum and in this case the over-voltage factor S can be expressed by

$$S = \frac{\text{Peak value of over-voltage}}{\text{Peak value of system voltage}} = K_m \frac{f_p}{f}$$

Parameters influencing the level of switching over-voltages

Magnetic efficiency K_m

The curve in Figure 17.4 represents the hysteresis cycle of the magnetic circuit in a transformer. K_m is a function of the ratio S/s or the difference between B_N (normal flux density in service) and B_R (residual flux density).

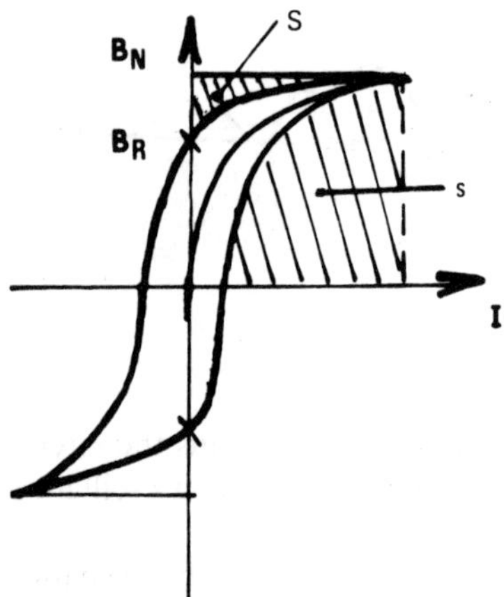

Figure 17.4.

Since the introduction of grain oriented sheet the normal flux densities have increased and the difference $B_N - B_R$ has reduced considerably (with mitred joints cut at 45°) and with it the value of K_m.

The magnitude of over-voltages is as follows:

(a) For a 10 MVA transformer with f_p about 400 Hz and $K_m = 0.3$ the over-voltage factor is

$$0.3 \times \frac{400}{50} = 2.4.$$

If the system is at 24 kV the peak value

$$= 24 \times \frac{\sqrt{2}}{\sqrt{3}} \times 2.4$$

$$= 47 \text{ kV, an acceptable value.}$$

(b) When a reactor or a transformer loaded by a reactor is disconnected, K_m approaches 1, f_p = 400 Hz and the over-voltage factor = 8 and the peak voltage 157 kV (for a system voltage of 24 kV).

This value is unacceptable and requires preventative measures: protection by arresters or the choice of circuit-breaker without over-voltage (hence without current chopping).

Breakdown strength of the gap between opening contacts of a circuit-breaker

The over-voltage above affects not only the transformer but also the intercontact space of the circuit-breaker. The re-strikes that occur in the intercontact space of the circuit-breaker delay the current break and reduce the value of the current finally broken with the result that the resulting over-voltage is restricted.

Special circuit-breakers for disconnecting reactor (or transformers loaded by reactors)

It has been seen that for disconnecting transformers on no-load, the term K_m alone can be sufficient to limit the over-voltages without questioning the choice of circuit-breaker.

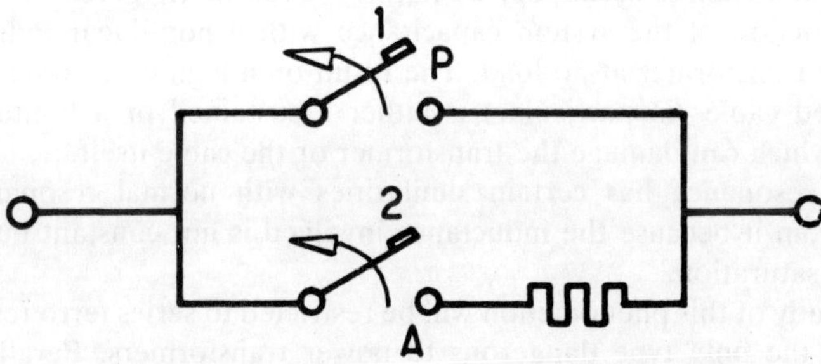

Figure 17.5.

When high voltage reactors have to be connected and disconnected frequently from the system, the use of special circuit-breakers with auxiliary breaking chambers fitted with non linear resistances is called for (see Figure 17.5). The operation is carried out in two stages: first the main chamber P is opened (1), then the auxiliary chamber A (2).

Conventional form of a switching over-voltage

Research has shown that the form 250/2500 has the maximum severity for distances in air. However, for transformers of rated voltages greater than or equal to 300 kV (the only transformers for which this test was intended) a similar form is applied with wide tolerances because of a slightly different behaviour.

Over-voltages arising from switching-in

When the supply system is of a high power relative to the transformer to be connected, it has a tendency to impose its own regime, and over-voltages due to switching in are hardly ever encountered.

However, if the supply system is of limited power the switching-in current represents a disturbance, and the return to a stable condition can only occur after there have been a series of transient voltage oscillations at the transformer terminals.

The over-voltage factor in this case cannot exceed

$$1.8 \times 1.5\frac{\sqrt{2}}{\sqrt{3}} = 2.2 \text{ which is an acceptable value.}$$

FERRO RESONANCE

This phenomenon is oscillatory by nature. It can be triggered off through the interaction of the system capacitance with a non-linear inductance, such as a transformer at no-load. The result on a high capacitance system (armoured cables in particular) is either a sustained or a limited over-voltage which can damage the transformer or the cable itself.

Ferro resonance has certain similarities with normal resonance but differs from it because the inductance involved is not constant but varies with the saturation.

The study of this phenomenon will be restricted to series ferro resonance which is the only type dangerous to power transformers. Parallel ferro resonances, for example, are produced with voltage transformers.

Resonance in a series LC circuit

When a circuit such as that shown in Figure 17.6 is supplied by an alternating current of frequency f (angular frequency $\omega = 2\pi f$) and the condition of resonance $L\omega = 1/C\omega$ is obtained, resonance occurs characterized by the fact that the impedance between the terminals a and b is at a minimum (almost zero if the losses are negligible).

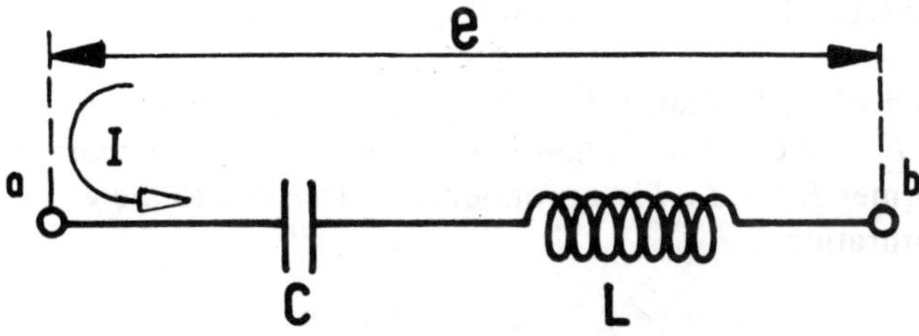

Figure 17.6.

Because Z is very small, even a very low voltage, e, applied to the terminals a and b will maintain a very high current of r.m.s. value I. This current creates voltages at L and C which are each of equal r.m.s. values (because the instantaneous voltages in the inductance and the capacitance are in direct phase opposition).

$$L\omega I = \frac{I}{C\omega} = E$$

E will be large compared with e, and there will be an over-voltage, with the over-voltage factor limited only by the losses in the circuit.

This phenomenon can start if the fundamental frequency of the system, or a harmonic, coincides with the fundamental frequency of the circuit:

$$f = \frac{1}{2\pi \sqrt{LC}}$$

Series ferro resonance

If to the circuit shown in Figure 17.6, a magnetic circuit is added to the coil, so that the inductance L is affected by saturation, making L dependent on voltage and frequency, then we have the situation shown in Figure 17.7.

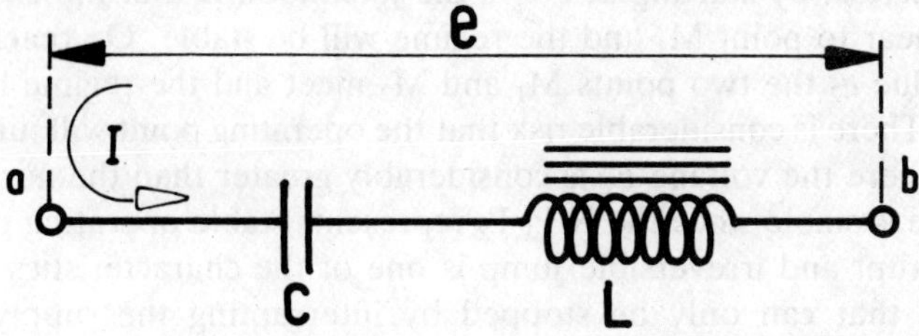

Figure 17.7.

If a current I flows through the circuit from a to b, and the losses are ignored,

$$e = Z_{ab} \times I = (L\omega - 1/C\omega)I$$

The two voltages E_c and E_L are always in opposition but are only equal if $L\omega = 1/C\omega$.

If a graph of the voltages $U = f(I)$ is drawn (as in Figure 17.8), it is found that $E_c = I/C\omega$ is a straight line, while the no-load characteristic of the transformer $E_L = L\omega I$ is a pronounced curve due to the variation of L with the saturation.

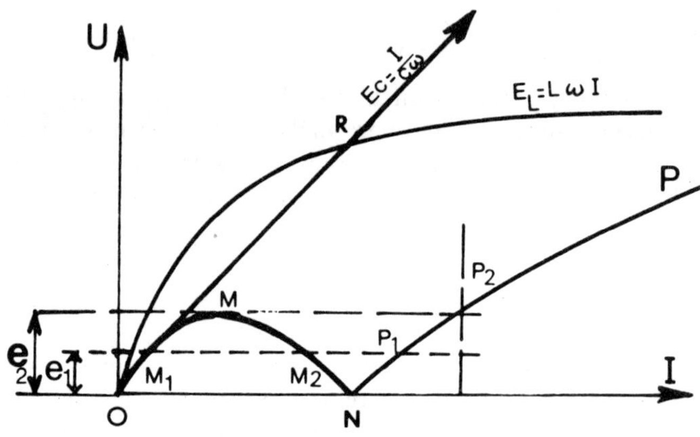

Figure 17.8.

At first glance only the point R is likely to cause a resonance due to $L\omega = 1/C\omega$.

However if the curve OMNP is drawn, representing the voltage at the terminals $e = (L\omega - 1/C\omega)I$, it is found that at a voltage e, there are three possible operating points M_1, M_2, and P_1 on the curve. Points M_1 and P_1 represent stable conditions while point M_2 represents an unstable condition as any variation of e near it causes a variation of I which does not compensate but accentuates the initial variation.

Nevertheless, by starting at $e = 0$ the likelihood is that the circuit will function near to point M_1 and the regime will be stable. On reaching the critical value e_2 the two points M_1 and M_2 meet and the regime becomes unstable. There is considerable risk that the operating point will jump from M to P_2 where the voltage E_c is considerably greater than the normal and the regime is stable since the N P_1 P_2 represents stable operating points.

This abrupt and irreversible jump is one of the characteristics of ferro resonance that can only be stopped by interrupting the supply to the circuit.

As before, the phenomenon can start at the nominal frequency or at a harmonic (and often at a sub-harmonic), but sometimes it changes from one frequency and fixes on another, making it particularly mysterious.

An exact mathematical analysis is therefore very difficult.

Analysis of some cases of ferro resonance

A 30 MVA, 63 kV transformer supplied by three single-phase cables of 150 mm^2 cross-section and 1800 m length was subjected to ferro resonance following a mechanical failure of a circuit-breaker, in which one phase was not closed during switching in at no-load.

The capacitances of each single-phase cable are about 0.25 μF (0.14 μF/km) and can be represented by three capacitors C_A C_B C_C whose impedance at 50 Hz is $Z_c = 1/C\omega = 12\,700\ \Omega$ (see Figure 17.9).

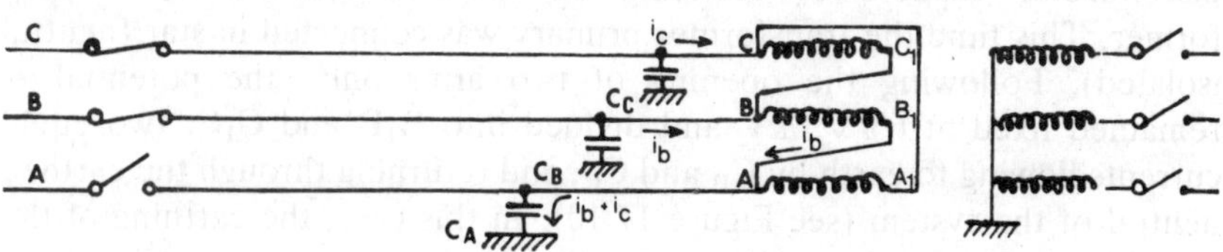

Figure 17.9.

Following the failure of phase A to close, the terminals B and C have their potential fixed by the phase-earth voltage of the system, and the potential of terminal A will be fixed by the currents i_c and i_b which flow through CC_1B and BB_1A and return by C_A, C_B and C_C (or by the earth and the neutral of the system if the latter is at earth).

These conditions come together to create a circuit in which an inductance, that of column A, and a capacitance C_A make resonance possible.

The magnetizing power demanded by the transformer is about 1 per cent, or $0.01 \times 30\ \text{MVA} = 0.3\ \text{MVA}$ (or its magnetizing current I_O is about 1 per cent of I_N). The equivalent impedance per phase in the equivalent circuit to the transformer at no-load is

$$Z_t = \frac{U^2}{P} = \frac{63^2}{0.3} = 13\,200\ \Omega.$$

It can be seen that the impedances Z_C and Z_T are dangerously near (4 per cent) to one another and that a weak disturbance could cause ferro resonance to start.

This situation was unfortunately reproduced involuntarily during a test and damaged a transformer. After repair of the circuit-breaker and the transformer, the installation was returned to service, and for safety the circuit-breaker was brought to an enclosure 50 m from the transformer. This eliminated the risk presented by the capacitance of the cable if the circuit-breaker should fail again.

An incident of the same type occurred in a transformer of about 15 MVA in similar conditions, following the failure to open of one arm of

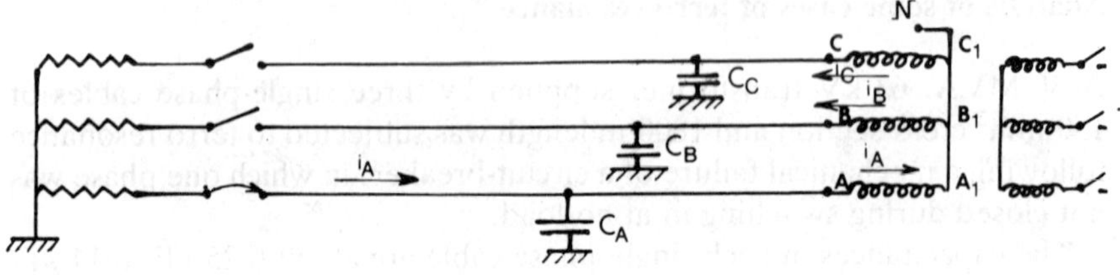

Figure 17.10.

disconnector capable of breaking the no-load current of a transformer. This time the transformer primary was connected in star (neutral
isolated). Following the opening of two arms only, the potential A
remained fixed at $63/\sqrt{3}$ kV and divided into B_1B and C_1C, two equal
currents flowing to earth by C_B and C_C, and returning through the earthed
neutral of the system (see Figure 17.10). In this case, the earthing of the
transformer neutral through a current limiting impedance was sufficient to
reduce the ferro resonance.

An unusual example

In a very high voltage system (525/230 kV) a bank of three single-phase
auto-transformers was taken out of circuit by the opening of a 230 kV
circuit-breaker close to the transformer and by the opening of a 525 kV
circuit-breaker some 30 km away. One of the three phases was subjected
to ferro resonance, started by the current induced capacitively by a second
525 kV line parallel to the first and about 30 metres away (see Figure
17.11). Neither of the 525 kV lines was transposed.

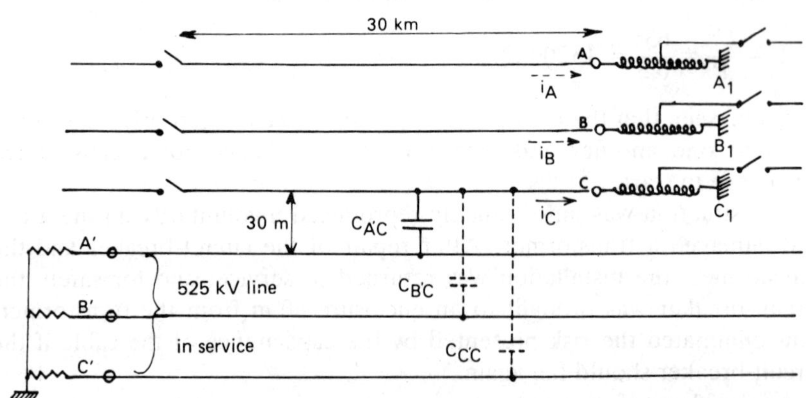

Figure 17.11.

Between each phase A′ B′ C′ of the line still in service and each phase A B C of the line out of service but connected to the auto-transformer, there were multiple unequal capacitances which by influence caused currents I_A, I_B and I_C to flow to earth through the auto-transformers, the most important of which was I_C (only the phase C was involved in this incident).

The circuit A′$C_{A'C}$ C C was thus closed by the earth and consisted of a capacitance and an inductance in series. Because of the very high phase voltage $525/\sqrt{3}$ kV (and probably the very low current I_O of an auto-transformer, often less than 0.03 I_N) the phenomenon of ferro resonance was able to start at the frequency of 60 Hz and maintain itself (sometimes at 60 Hz, sometimes at 20 Hz).

Up to this point the tertiary winding on each unit had been left disconnected, but after connecting these in delta the affected unit was loaded by the other two and the phenomenon ceased to appear.

Conditions favourable to the appearance of ferro resonance

Ferro resonance may occur when:

the transformer is on no-load (even an active low load, less than one-tenth of the rated power is sufficient to prevent this phenomenon by absorbing the energy),
a large capacitance exists in line or cable which can be placed in series with at least one transformer terminal, the others remaining at the system potential,
an unbalanced situation is created by a single-phase trip out, a break in a conductor, or blowing of one or two fuses.

Various remedies can be deduced from these possibilities:

reduce the length of line between the transformer terminals and the section circuit-breaker or fuses,
avoid non simultaneous switching,
earth the neutral point directly or through a resistor limiting the single-phase fault current.

TRANSMISSION OF OVER-VOLTAGES FROM PRIMARY TO SECONDARY

Frequently over-voltages applied to one transformer winding will be transmitted to the other winding, and depending on the conditions this transmission may harm the winding concerned or the equipment connected to it.

Inductive transmission

The operating principle of a transformer is based on making use of mutual induction between two windings. Tight coupling between two windings causes an almost perfect reproduction in one winding of the phenomenon occurring in the other. So far as industrial frequencies are concerned, the purpose of the magnetic core is to increase this coupling, the leakage flux being the only departure from perfection. Consequently all phenomena of an impulsive or oscillatory nature applied to one winding are reproduced faithfully in the other winding in the ratio of the numbers of turns (transformation ratio).

However, the transformation ratio alone cannot account for all the phenomena of internal oscillation either because of the steepness of the front or because of the attenuation due to the presence of the magnetic circuit whose laminated construction responds only to relatively low frequencies. So a correction factor must be introduced.

For lightning over-voltages reaching a transformer whose LV side is only lightly loaded, the coefficient is generally less than 1.3, but in exceptional cases it can be as high as 1.8. For switching over-voltages in the same conditions, the coefficient rarely exceeds 1.8.

Take, as an example, a 20000/400 volts Dyn 11 transformer. If a lightning impulse of 95 kV peak and 1.2/50 wave shape is applied to one of the phases of the HV winding, this very short unidirectional impulse (50 μs to half amplitude) on the primary winding is transmitted to the secondary in more or less the same form with a peak value less than $1.8 \times 95/87 = 2$ kV (transformation ratio $20000/400 \times \sqrt{3} = 87$).

If, however, an over-voltage wave due to switching of 250/2500 wave shape and amplitude $2.4 \times 20\sqrt{2} = 83$ kV peak, is applied to the primary, the secondary impulse would conform more to the model form and would have peak value $1.8 \times 83/87 = 1.7$ kV.

Generally, HV and LV systems are coordinated in such a way that these transmitted over-voltages do not have any harmful effects on the equipment. The construction of transformers is such that a considerable coefficient of safety is conferred with respect to over-voltages transmitted inductively, since the configuration of the LV windings imposed by mechanical or thermal considerations provides insulating distances above those strictly required by electrical considerations.

Capacitive transmission

This second effect of transmission is not related to the operating principle of transformers but to the arrangement of the windings.

In a transformer with concentric windings, each of the windings has a distributed capacitance relative to its neighbouring windings and/or rela-

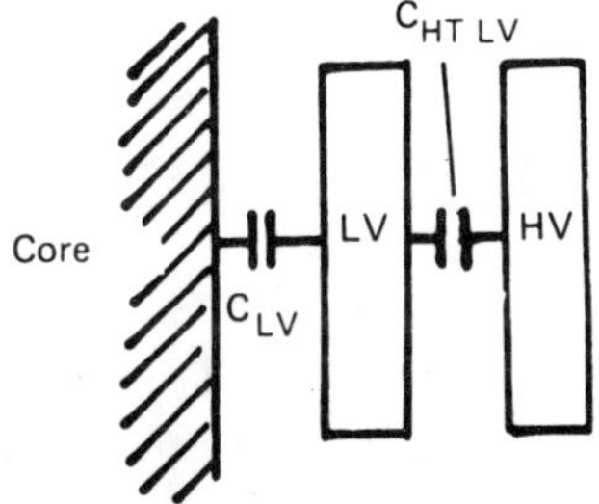

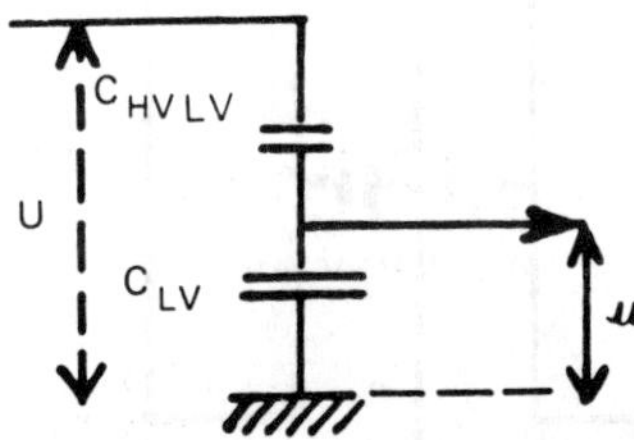

Figure 17.12.   Figure 17.13.

tive to the magnetic circuit, which will be represented by lumped capacitances C_{LV} (between LV and core) and C_{HVLV} (between HV and LV). At first sight any voltage applied to the HV winding could be transmitted capacitively to the LV winding and the system of capacitances would behave as a capacitance divider such that the voltage U applied to the HV side would be transmitted to the LV side in the ratio

$$\frac{u}{U} = \frac{C_{HVLV}}{C_{HVLV} + C_{LV}}$$

The construction is such that the LV winding is usually near the core and relatively far from the HV winding, and consequently C_{LV} is considerably greater (2–8 times) than C_{HVLV} which reduces u/U to a value between $\frac{1}{4}$ and $\frac{1}{9}$.

Moreover it is rare for the potential of the HV winding to be high throughout. Often the neutral point will be at zero potential (earth) while the HV terminal is subjected to over-voltage. In this case the representation by lumped capacitances is incorrect, and distributed capacitances should be considered and a calculation made by integration, which in the average case would reduce the factor by one third, bringing u/U to between $\frac{1}{12}$ and $\frac{1}{27}$.

To allow for possible superimposition of the impulse wave on a peak of the system voltage at 50 Hz, these factors should be increased by 1.15 to raise them to $\frac{1}{10}$ and $\frac{1}{24}$ respectively.

However, when an HV winding is struck by a very steep fronted lightning impulse wave, the distribution of this wave along the HV winding is never linear and tends to give rise to oscillations. The earlier method of considering the problem is then no longer valid.

The transmission factor remains related to the relative magnitudes of the winding capacitances but it can vary within wide limits depending on the circumstances.

It should be noted that these transmissions are considerably attenuated and even eliminated if at least one point of the LV winding is connected to a potential near to earth. This is why the presence of voltage limiting

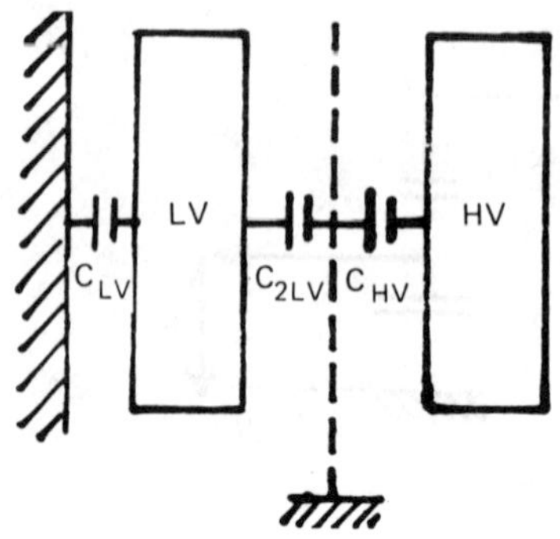

Figure 17.14.

devices on each phase or on the neutral point, or the solid earthing of the neutral point, are to be recommended.

It should also be noted that the presence of lines connected to the LV side often adds capacitances in parallel with C_{LV} which increase their apparent value and play an important role in reducing transmitted over-voltage.

In spite of all this, in some installations transmitted over-voltages can have a harmful effect on the LV winding of a transformer, and also the equipment connected to it, without it being possible to impose earth potential at the neutral point.

In these cases it is recommended that special transformers be used, fitted with electrostatic screens between windings (Figure 17.14). This screen consists of a sheet of metal foil (or a wire grid of conductors) placed between the HV and LV windings, insulated from each and connected directly to earth, which allows all the electrical charge transmitted by C_{HV} to flow to earth. This short-circuits the lower arm of the voltage divider and consequently eliminates all capacitive transmission. (Clearly these electrostatic screens must be designed so that they do not form a short-circuited turn between the two windings.)

Thus, it can be seen that over-voltages transmitted from the HV winding to the LV winding remain within reasonable limits, that they can be overcome by voltage limiting devices and that the use of electrostatic screens is only effective against capacitive transmissions and is only justified in special cases.

WITHSTANDING IMPULSE, LIGHTNING AND SWITCHING WAVES

Lightning over-voltages

Lightning impulse has a very short duration to half amplitude, and so its destructive effects are limited by comparison with those of a stress at 50 Hz of the same peak value.

The impulse factor is given by

$$\frac{\text{impulse strength}}{\text{power frequency strength} \times \sqrt{2}}$$

and has a magnitude of about 1.8.

When the field is not completely uniform, partial discharges (analogous to corona effect in air) have the effect of making the electric field more uniform and retarding complete breakdown.

However, a transformer designer endeavours to obtain uniform configurations of field, in which the breakdown voltage depends only on the amplitude of the incident wave and not on its steepness.

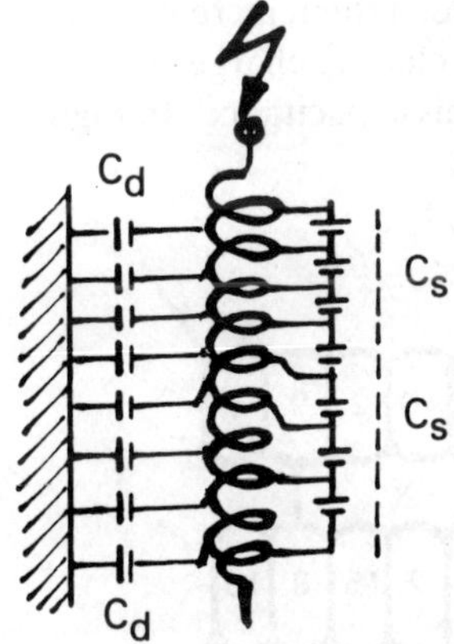

Figure 17.15.

A transformer winding struck by an impulse behaves like a system of distributed constants consisting of series inductances (each turn is an inductance), series capacitances (capacitance of a turn to its neighbour) and shunt capacitances (capacitance of a turn to earth). It can be shown that the initial distribution of impulse voltage is not linear but dependent on the coefficient $\alpha = \sqrt{C_d/C_s}$.

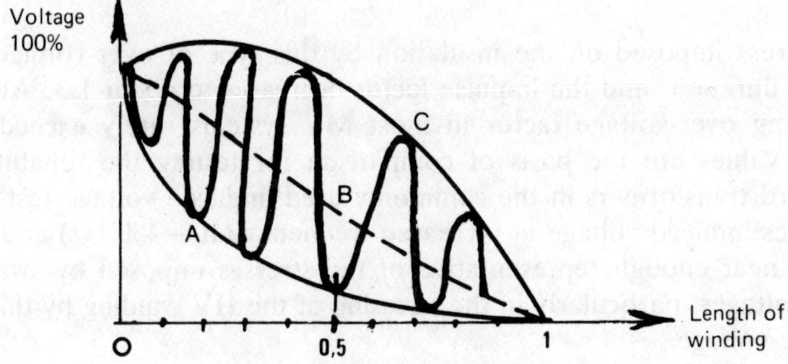

Figure 17.16.

The curvature of curve A in Figure 17.16, which represents the initial distribution of voltages along the winding that has been struck by an impulse, will increase as α increases.

The steady state B will be reached after a series of oscillations between curves A and C which represent the envelope.

The aim would be to obtain an initial distribution such as B, eliminating all oscillation (non-resonant winding) which would assume $\alpha = 0$. This can only be approached by reducing C_d and increasing C_s as much as possible.

For transformers of rated voltage above 245 kV, a type of winding known as an 'interleaved winding' is frequently used in which turns are brought together out of sequence and not in the order 1, 2–2, 3 etc. This does not alter the physical capacitance between two physically adjacent conductors but because of the higher voltage between them increases (here in the approximate ratio $8^2 = 64$) the amount of electric charge transmitted in the capacitance C_s and hence the effect of this capacitance. In Figure 17.17 the increase is approximately $8^2 = 64$ times.

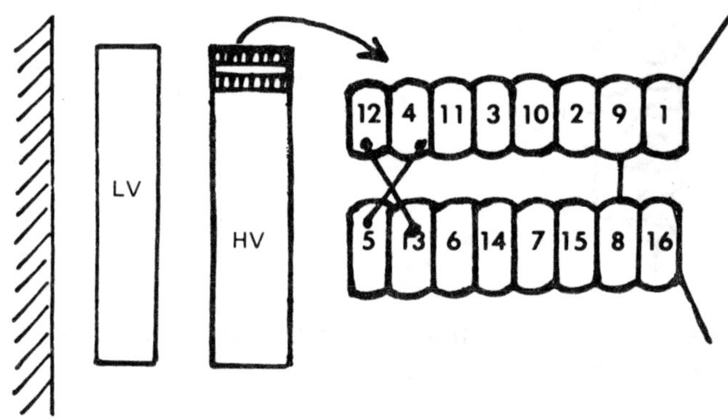

Figure 17.17.

Switching over-voltages

The stress imposed on the insulation by this type of over-voltage is of longer duration, and the impulse factor decreases to about 1.3. Also the switching over-voltage factor in most MV systems rarely exceeds 2.4. These values are the basis of comparison for testing the reliability of standard transformers in the commonly used 'induced voltage test'. This test uses induced voltage at increased frequency (100–400 Hz) and 2 U_n, and is near enough representative of the stresses imposed by switching over-voltages, particularly in the stressing of the HV winding by this type of over-voltage.

It has already been noted that the fundamental wave of a switching over-voltage 250/2500 approximates to a frequency of oscillation of 200 Hz. The phenomenon is therefore slow, the effect on the factor α becomes negligible, and a uniform distribution is obtained represented by the straight line B in Figure 17.16. This adds weight to the validity of the test by induced voltage at $2\,U_n$.

PROTECTION OF TRANSFORMERS AGAINST OVER-VOLTAGES

Substations and transformers are usually protected by two types of appliances: arresters with non-linear resistors, and spark gaps.

Arresters with non-linear resistance

Their role is to restrict over-voltages, and they also limit the follow through current. Their characteristics are governed in France by standard specification UTE C 65 100.

Their level of protection is specified by:
spark-over voltage on a full wave 1.2/50,
residual voltage at the standard discharge current (5–10 or 20 kA),
front of wave spark-over,
possible spark-over voltage for switching over-voltage 250/2500.

These values are given by the supplier (see Table 17.2 and Figure 17.18).
Effective protection by arresters requires that they be placed next to (i.e. within a few metres of) the equipment they are protecting.

Spark gaps

This device can lop the peak of some over-voltages but neither restricts nor interrupts the resulting current. Breakdown of the gap results in a single phase to earth fault, or a fault between phases when there is simultaneous breakdown of several spark gaps. This can only be cleared by the action of relays and circuit-breakers.
Table 17.3 provides some characteristics for the setting of spark gaps used in a large number of medium to high tension systems.

Comparison between these two methods of protection

Spark gaps vary widely in their flashover voltages and so their flashover voltage is defined as that which gives 50 per cent flashover. They are sensitive to the steepness of the incident wave.

This can be explained by the spark gaps having a time lag to flashover which, if the gap is set at 350 mm varies from 0.5 μs for a wave front of 1000 kV/μs to 3 μs for a wave front of 300 kV/μs, passing through 1 μs for 600 kV/μs.

In Figure 17.19 is shown a series of results of experiments carried out on a spark gap adjusted to 350 mm and subjected to impulse waves 1.2/50 of increasing value. The successive breakdown points describe the voltage/time curve for such a gap.

Another cause for imprecision is that during heavy rainfall, which often occurs during storms when spark gaps are most required to fulfil their role, a big increase in the variation of flashover voltage is found during the first half hour.

EFFECT OF INSULATION ON COST

The insulation level chosen for a transformer has an effect on its purchase price. These differences can affect the choice of level of insulation, higher to avoid the purchase of arresters, or lower with the use of arresters. As a general rule, arresters are not commonly used by MV consumers (17.5–20 or 24 kV).

Table 17.1 Insulation level

Rated power (MVA)	Lightning impulse (kV peak)	Price
1	95	100
	125	102
40–100	900	100
	1050	104

For a big MV system (72.5–90 kV) the saving made by not installing arresters can often cover the cost of damage.

Usually at 245–420 kV, the installation of arresters is considered to be economic particularly if the risk of lightning strikes is high.

APPENDIX

Characteristics of arresters Type HML

Table 17.2

Rated voltage:	4–150 kV
Frequency:	40–60 Hz
Rated discharge current, peak value:	10 kA
Maximum discharge current, peak value:	100 kA
Resistance to current waves of long duration, peak value:	2000 µs, 600A
Discharge class in long duration wave:	Class 2, IEC
Pressure relief device Class A:	40 kA minimum
Installation altitude:	not exceeding 1000 m*
Installation altitude above 1000 m:	on request

Characteristics guaranteed

Type HML	Maximum supply voltage permitted at the terminals of the arrester	Minimum flashover voltage at supply frequencies	Maximum voltage at 100% flashover with impulse wave 1.2/50 µs	Maximum flashover voltage at switching over voltages	Maximum voltage of flashover on the wave front**	Maximum value of residual voltage with wave of 8/20/µs and impulse current of		
						5000A (Peak value kV)	10000A (Peak value kV)	20000A (Peak value kV)
	(r.m.s. value kV)	(r.m.s. value kV)	(Peak value kV)	(Peak value kV)	(Peak value kV)			
4	4	7	13	14	16	10	11	12
7.5	7.5	13	24	26	30	15	17	18
12	12	20	37	40	46	23	25	27
15	15	25	47	50	59	30	32	35
18	18	30	54	58	68	35	38	41
25	25	43	75	80	94	49	53	57
30	30	50	87	95	109	59	63	68
37	37	63	105	110	130	73	78	84
40	40	68	110	120	135	79	84	91
55	55	94	150	160	185	107	115	124
60	60	100	160	170	200	117	126	136
75	75	125	200	215	250	146	157	170
90	90	150	230	235	290	175	190	205
100	100	170	250	255	310	195	210	227
110	110	185	275	280	340	215	231	250
120	120	205	290	300	360	234	252	272
135	135	230	325	340	400	264	285	307
150	150	255	350	375	435	293	315	340

* On all orders for arresters specify the installation altitude.
** The steepness of the wavefront corresponds to 100 kV/µs per 12 kV of the rated voltage of the arrester up to 1200 kV/µs maximum.

Dimensions of arresters Type HML

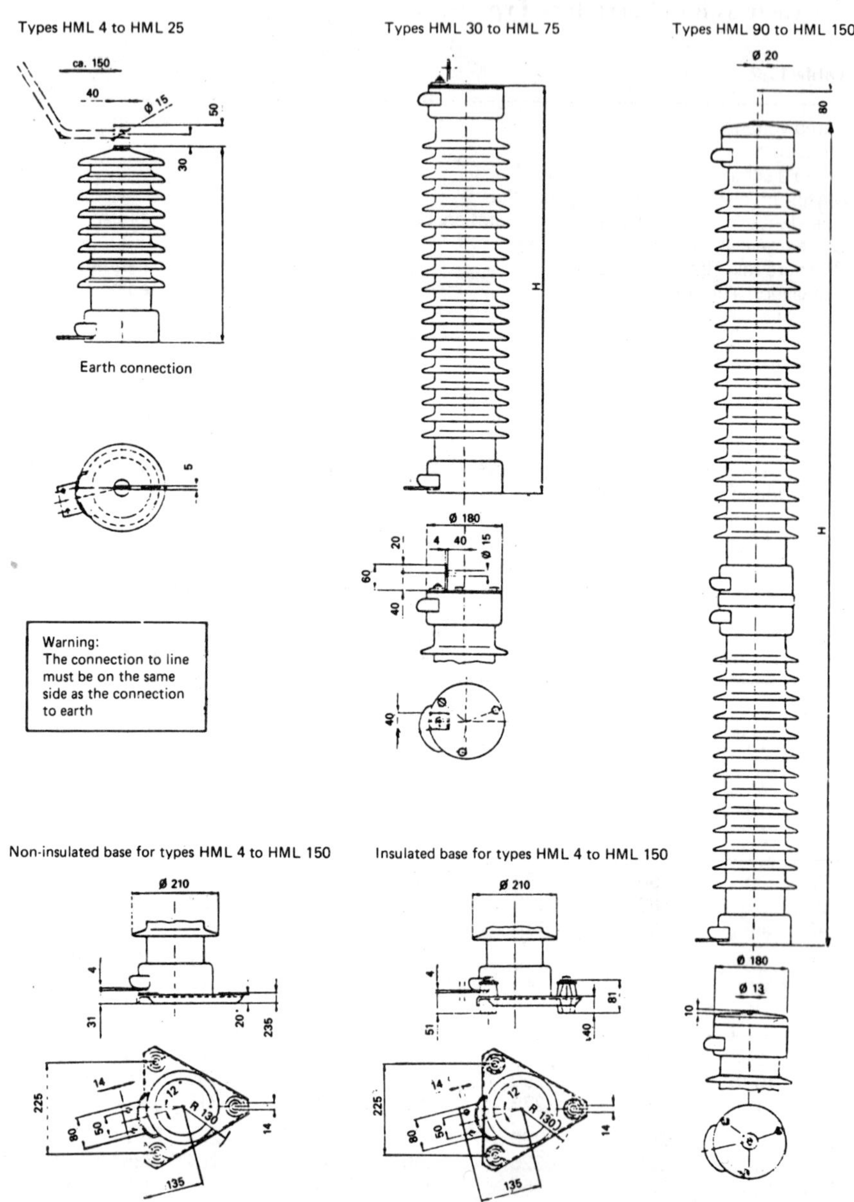

Figure 17.18.

Setting of gaps for protection

Table 17.3

Supply voltage $(kV_{r.m.s.})$	Rated test level of equipment		Substation spark gaps		Transformer spark gaps		Distance recommended between metallic parts phase/earth in air $(mm)*$
	Impulse (kV_{peak})	50 Hz $(kV_{r.m.s.})$	Gap setting (mm)	Voltage giving 50% impulse flashover (kV_{peak})	Setting (mm)	Voltage giving 50% impulse flashover (kV_{peak})	
12	75	28	25	60	In this range of voltages		120
17.5 – 23	95	38	40	75	transformer spark gaps are not		160
24	125	50	70	95	used frequently, or used with		220
36	170	70	120	135	settings as for substations		320
52	250	95	200	200	270	230	480
72.5	325	140	280	250	350	300	630
100	450	185	400	330	480	420	900
170	750	325	650	500	820	620	1500

* These values are given for guidance only so that any installation can have capacity to resist impulse at least as good as the equipment used. Minimum distances are not laid down if the equipment is subject to an impulse test, as the distances laid down can cause difficulties in the design of the equipment, increase its cost and restrain progress. The impulse test itself is sufficient to prove that the conditions for resisting impulse are adequate.

Experimental characteristics of a rod gap

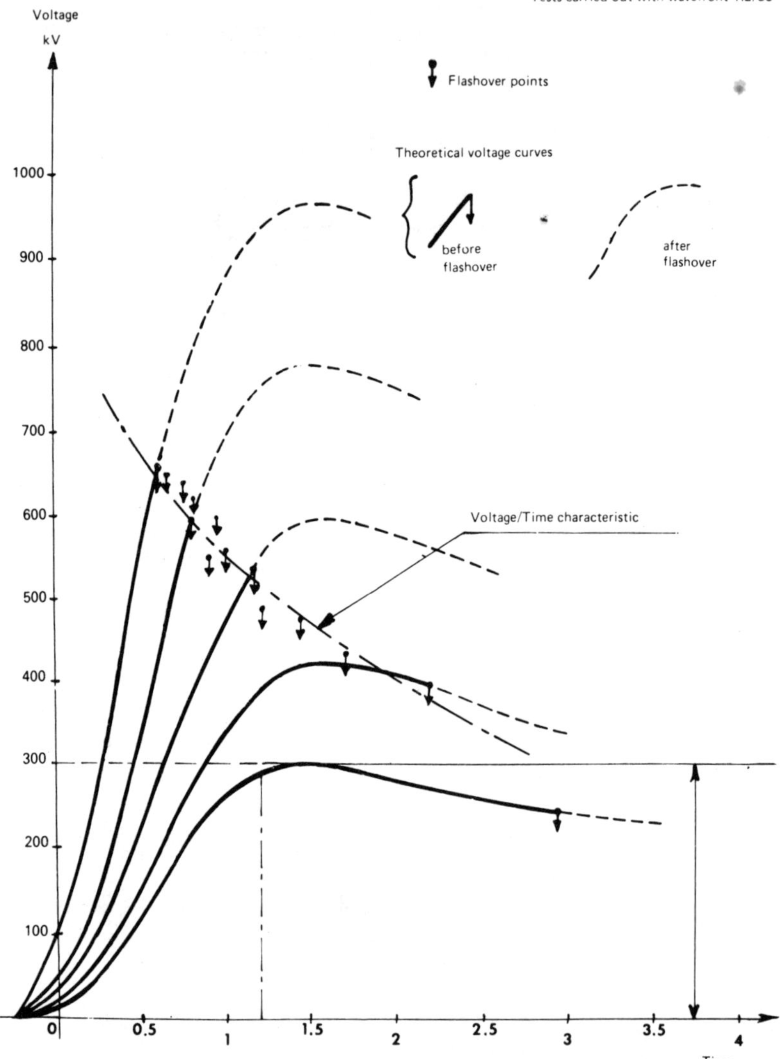

Figure 17.19. Voltage-time characteristic for a rod gap (experimental)

18 MV/LV transformers

MV/LV TRANSFORMER TECHNOLOGY

There are many MV/LV transformers in the world and so they merit a chapter to themselves. We consider here oil-insulated transformers. (Dry type transformers are not considered in this book).

Figure 18.1. 630 kVA distribution transformer, 20 000/400 V

Line transformers

These transformers of standard design consist of core and windings, tank and cooling, and accessories.

Core and windings

The low-voltage winding consists of a multi-layered cylinder covering the whole height of the winding and having a sufficient number of turns (generally 10–30) to obtain the required voltage.

The high-voltage winding is wound in layers with enamelled round wire. The tappings for off-circuit regulation are taken out from the external layers.

Tank and cooling

The tank is rectangular in shape and consists of corrugated panels which dissipate the losses.

Equipment and accessories

All the usual equipment and accessories can be mounted on the transformers. The specification often depends on the requirements of the user.

Transformers of balanced flux in the magnetic circuit

They differ from line transformers by their magnetic circuits and their tanks. The magnetic circuit consists of three equi-spaced legs and two triangular shaped yokes.

The tank takes the shape of the core and windings, and the cooling is obtained by banks of radiators. This type of transformer has not been manufactured since 1979.

MV/LV SUBSTATIONS

The first prefabricated substations appeared on the market in the 1950s. Their advantages include:

permanent and complete security,
economic construction benefitting from mass production,

reduction of site erection time,
factory tests for reliability,
reduction of size, and hence the cost of the building housing the equipment.

Summary of characteristics

Rated voltage:

MV = 5.5, 10. 17.5, 23, 24 kV,
LV = 230, 400, 500, 600 kV

Transformer rated power: 800–2500 kVA
Standardized short-circuit currents for the EDF 20 kV systems: 8–12.5 kA.

Layout

MV/LV substation layouts are closely related to the guarantees given regarding continuity of electrical supply. This continuity obviously depends on the high voltage, 63–90 or 225 kV, used (see Figure 18.2). The latter is sufficiently interconnected, however, for there to be rarely any doubt about the permanence of supply at medium voltage.

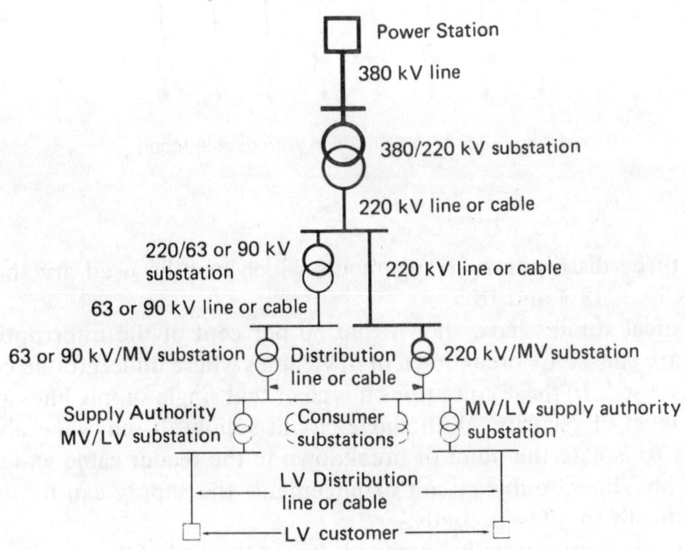

Figure 18.2. Electrical power distribution system

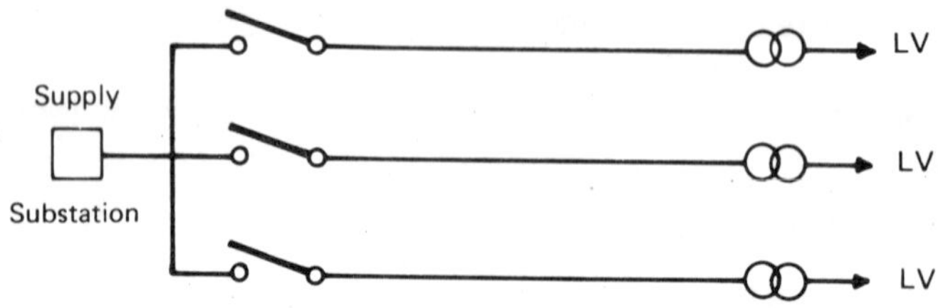

Figure 18.3. MV distribution with single feeder

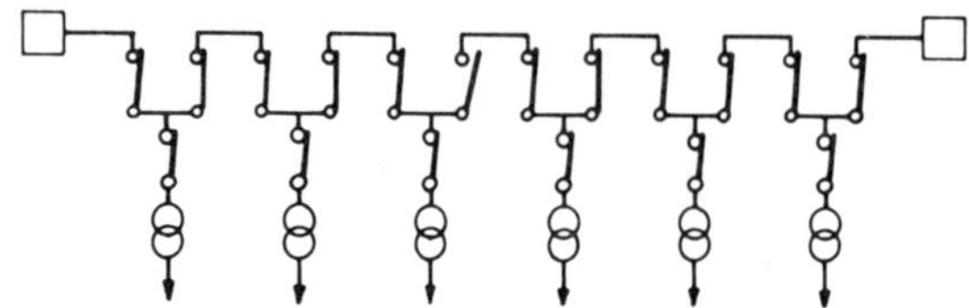

Figure 18.4. MV distribution from two substations using single split feeder

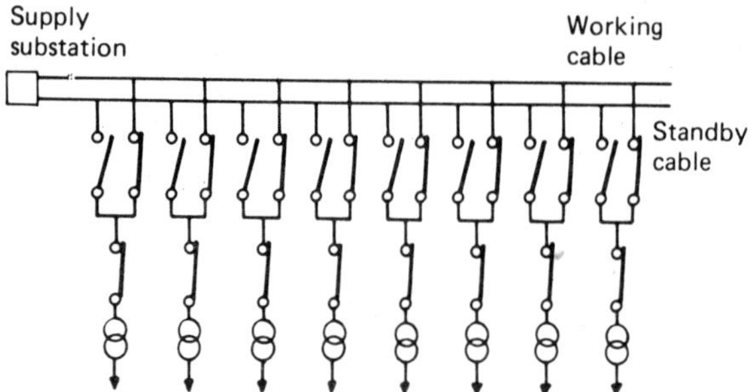

Figure 18.5. MV distribution with double feeder

The three distribution arrangements which may be used are shown in Figures 18.3, 18.4 and 18.5.

Statistical studies have shown that 80 per cent of the interruptions to supply are caused by breakdown of MV cables where underground connections are used. In these conditions it is clear that single supply lines give the lowest level of security. With single-circuit supply, it may take about 30 minutes to isolate the point of breakdown in the feeder cable and restore the supply. In a double-circuit arrangement, the supply can be restored automatically in a few seconds.

The LV system is usually overhead. It is at the end of the supply system, where it is usually very reliable. In any case it is easily repaired.

Security, reliability, maintenance

The standards which govern MV/LV substations specify the level of security in various circumstances and require the manufacturer to design and construct the equipment accordingly. The maintenance of the components is carried out in accordance with the recommendations made by the manufacturers. In general the number of switching operations on the MV side is low, and interference with the supply very infrequent.

Making-off cable sealing boxes is often the most difficult operating problem. With normal size substations the dimensions are well established at 700 mm wide and 1000–1100 mm depth. Smaller substations (width 500 mm or less) require very special provisions to be made.

CONNECTIONS

There are various methods of connecting the MV and LV sides, for example:

Medium voltage
   dry cables,
   single-phase oil filled cables,
   three-phase oil filled cables,
   overhead conductors.

Low voltage
   bars when the board of LV substation is near to the transformer,
   cables when the LV substation is distant.

The criteria which influence the choice between these different solutions are:
   the level of security against direct contact,
   the risk of fire (cables being the safest in this respect),
   the rapidity of disconnection which in the case of cable boxes leads to the use of disconnecting chambers, and with dry cables the use of terminations.

The connections have standard dimensions particularly when terminations are used. The LV exits are often protected by covers.

PROTECTION OF CONSUMERS' SUBSTATION TRANSFORMERS

The protection of transformers has been dealt with in previous chapters, but it is useful to return to the subject of the regulations governing the protection of consumer transformers.

Figure 18.6. Distribution transformer with protective cover for the LV terminals

The current standards contain a certain number of regulations of which the most important can be grouped under four headings:

Protection against indirect contact.
Protection against internal faults.
Protection against over-voltages.
Protection against excessive currents.

Protection against indirect contact

Only the regulations concerning the earthing of neutral points are considered here.

Earthing of neutral points on the supply side

There is no regulation in the French Standards for earthing of neutral points of primary windings in transformers, for two reasons. Firstly most MV/LV transformers in the power range 250–2500 kVA are provided with the medium-voltage winding connected in delta and therefore have no neutral. Secondly, the method of earthing a neutral has a direct effect on the fault currents in a system. They are the concern of the distributor, who does not want the consumers' installation to react on the characteristics of the system. In the same way, the consumer does not want a fault in the system to abnormally affect his equipment.

In most medium-voltage systems, the neutral is earthed in such a way as to limit (to 300 A in overhead systems and 1000 A in underground systems) the single-phase line to earth fault current, which corresponds to earthing of the neutral through a low value impedance.

Earthing of neutral points on the low voltage side

There are two principal arrangements for the LV neutral (see Table 18.1). In arrangement T, the neutral is connected solidly to earth or through a low value impedance, while in arrangement I, the neutral is isolated or connected to earth through a high value impedance. This second arrangement can only be considered satisfactory from the point of view of protection of transformers if it is noted that paragraph 443–2 of French standard NF C 13-200 adds in this case 'In principle over-voltage limitation equipment is provided on the input side of the equipment'.

Protection against internal fault

Relays to detect gas emission

Table 55A of the French standard advises, but does not insist, on the provision for transformers of 630–5000 kVA with a single contact relay, and for transformers above 5000 kVA with a two contact relay.

The effectiveness of this protection equipment has been demonstrated many times. It restricts damage by cutting off all power supplies as soon as gas is detected. It is sometimes overlooked however that some machines (synchronous and even asynchronous motors) can, while slowing down, behave as generators in converting their inertia back into energy. This makes it necessary for the gas detection relay to trip the circuit-breakers on the secondary side of the transformer as well.

Table 18.1 A comparison of earth connection arrangements

	Arrangement TT	*Arrangement TN*	*Arrangement IT*
Principle	The fault current flows in loop including the earthing connections of the neutral and the earthed metal equipment. 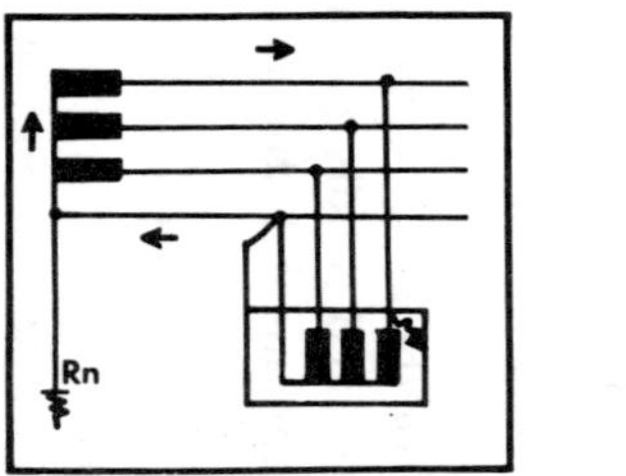	The fault current flows in a loop in the neutral conductor and becomes a phase-neutral short circuit current. 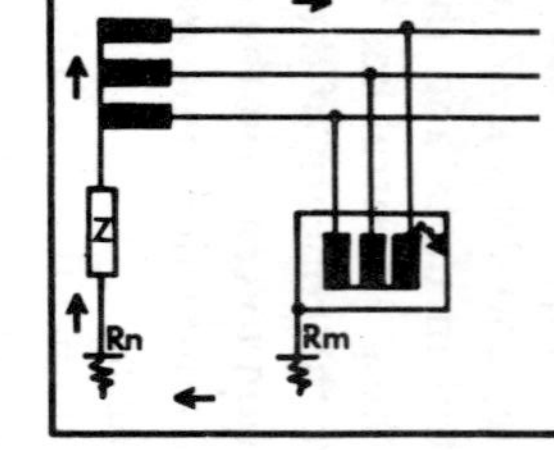	The current of the first fault is restricted to a value such that dangerous contact voltages do not arise.
Conditions	Differential tripping device must cut the supply as soon as the fault voltage exceeds U_L. This condition determines a relationship between the resistance of the earth connection of the equipment and the rated differential current of the device. $$R_m \, I_n \leqslant U_L$$ All the equipment protected by the same differential device must be connected to the same earth connection. The equipment simultaneously accessible must be connected to the same earth connection.	The fault current brings about the operation of the protection device against over-currents. This condition determines a relationship – which is that of protection against short circuits – between the impedance of the fault loop and the operating current of the protection device. $$Z_d/a \leqslant U_o \;^{(1)}$$ The protection conductor must be earthed properly so that its potential is as near as possible to that of earth. Precautions must be taken to avoid any break of the neutral conductor while it is being used for protection.	The first fault current does not cause any breaking device to operate, provided that the current is not greater than $$I_d \leqslant \frac{U_L}{R_m} \;^{(2)}$$ as a function of the earthing resistance. A permanent monitor of the insulation must indicate the first appearance of an insulation defect. In the event of a second fault the protection must be assured under the same conditions as in arrangement TN if all the equipment is interconnected or TT if not [7].

Use	Required for installations supplied directly by a low voltage public distribution system.	Usable only in installations supplied from a private substation.	Usable only in installations supplied by a private substation [3] and operated by a maintenance department.
Advantages & disadvantages	Disconnection on the first insulation fault. Requirment to install differential devices assuming protection against indirect contact [4].	Disconnection on the first insultation fault. Use of protection devices against over-currents to assure protection against indirect contact. Saving of one pole and one protection conductor in non terminal circuits of at least 10 mm^2 area and in fixed ducts. Requirements concerning the earthing of the protection conductor. Passage of the protection conductor in the same ducts as the live conductors of the corresponding circuits [5]. Frequent requirement to provide additional equipotential connections [6]	No disconnection on the first insulation fault and possibility to maintain continuity of service. Insultation monitoring requiring a maintenance department, enabling rapid elimination of faults. Non-distribution of neutral conductor, if not necessity to protect this conductor. Practical necessity of providing equipotential of equipment, if not installation of differential devices. Installation of over-voltage limiters. Limitation on extent of installations to limit intensity of first fault current.

Notes:

[1] The value of I_a is that which causes the operation of the protection device in a time corresponding, according to the safety curve, to the presumed contact voltage.

[2] The value of the first fault current I_d includes the leakage currents of the installation in normal service.

[3] Consumers substation is a substation supplied by an industrial installation in accordance with French standard NF C 13–200 or an autonomous source.

[4] In large installations, an operational selectivity can be provided between differential devices.

[5] This does not allow the use of the metal frame only as protection conductors, because of the increase in impedance of the fault loop that would arise, an increase which would impede the operation of the protection devices.

[6] Additional equipotential connections must be provided when condition[1] cannot be met, in order to impede the appearance of dangerous contact voltages.

[7] The determination of the tripping conditions of the second fault must be based on the value of the current of the double fault which is usually less than the first fault current in an installation connected as in TN.

Figure 18.7. Dry type distribution transformer

Protection against over-voltage

According to French standard NF C 13–200 this protection is required only
if the transformer is supplied by overhead lines.

The analysis of over-voltage phenomena (Chapter 15) has led us to draw
attention to the effect of the substation layout on the value of the over-
voltages, and remark that high risks occur when transformers are supplied
by an overhead/underground connection of a certain layout (reflection
phenomenon).

It should be emphasized that preference should be given to the use of
voltage limiting devices with non-linear resistance whose action is more
certain than protective air gaps and that, in difficult cases, the neutral, if
not already directly earthed, must also be provided with an arrester.

Protection against over-currents

This very wide range of phenomena may involve short circuits where
$I_{cc\ r.m.s.}$ is about 8–25 $I_{n\ r.m.s.}$, and $I_{n\ peak}$ is about 2.5 $I_{cc\ r.m.s.}$.

While a transformer should be capable of withstanding the thermal and mechanical effects of a dead short circuit at its terminals lasting 3 seconds, within that period of time the temperature of the copper windings can reach 250°C.

Obviously the repetition of such a thermal stress cannot be allowed, and it is advisable to provide equipment (isolators, circuit-breakers, fuses) capable of interrupting short-circuit currents in as short a time as possible.

A total time of interruption in the order of 0.2–0.5 s can be regarded as acceptable from the thermal point of view, but is too long to limit the electrodynamic forces on the windings which arise during a succession of decreasing assymmetric peaks, the first of which appears only 10 ms after the short circuit. Consequently the electrodynamic forces on a transformer are, in practice, only limited by its own impedance and that of the system.

Switching-in overcurrents ($I_{E\ peak} \simeq$ 8–10 $I_{N\ r.m.s.}$) appear as a series (50–500) of decreasing unidirectional impulses, repeating themselves 50 times per second. They do not have a sinusoidal wave form and are normally expressed by a factor K expressing the magnitude of the crest of the first impulse in terms of the effective value of the rated current $I_{E\ peak} = K I_{n\ r.m.s.}$ (where $K \simeq$ 8–10 for 2000 and 800 kVA).

The value of the coefficient n, where the threshold for operation of the tripping relay is nI_N, allowing switching-in without relay operation, will have only a distant relationship to K. The design of the relay makes it sensitive to the effective value of the current, because its average value, its peak value and its range are always graduated in effective values (using the conversions $I_{peak} = I_{r.m.s.} \sqrt{2}$ or $I_{mean} = I_{r.m.s.} \times 1.1$, which are only valid for sinosoidal conditions).

The result is that the ratio n/K can assume practically any value between 0.3 and 1.

Ability of a transformer to accept overload

Strictly, the curve $t = f(n)$, expressing the allowable duration of the overloads that a transformer can withstand, can only be established by the manufacturer who must take into account parameters such as size of transformer, ambient temperature, temperature rise before overload, maximum allowable temperature, duration and frequency of recurrence of overloads. Some of these can be assessed only arbitrarily.

This has required the preparation of a guide to loading (IEC 354) for overloads of long duration (half hour or more) and it is the subject of Chapter 6.

The behaviour of a transformer during heavy overloads (2–6 I_n) of short duration is sometimes not well understood, and this leads to extremes either of premature deterioration or of excessive caution.

In Chapter 16 it was shown that, assuming adiabatic conditions, the magnitude of the final temperature of a winding could be calculated from the formula

$$O_f = O_i + 0.128n^2t$$

from which $t = \dfrac{O_f - O_i}{0.128n^2}$

For a transformer of modern construction (800–2500 kVA) loaded continuously at its rated current, at an ambient temperature 30°C, the temperature of the hot spot in the windings reaches about 108°C. Peaks of 140°C for short periods are permissible.

Based on these assumptions the allowable duration of an overload can be expressed by

$$t = \frac{32}{0.128n^2} \simeq \frac{250}{n^2}$$

Table 18.2 shows values of t for assumed values of n.

Table 18.2

n	1.2	1.5	2	3	4	5	6
t	174	111	63	28	16	10	7

A simple calculation shows the effect of changing the initial assumptions. If the ambient temperature is 20°C instead of 30°C the new values of t can be obtained by multiplying the values in the table by

$$\frac{42}{32} \approx 1.3.$$

In the same way, if it is required to calculate the duration of overload from the cold condition (30°C) instead of the hot (108°C), the values of t in the table are multiplied by the factor

$$\frac{110}{32} \simeq 3.4.$$

These calculations, of course, only give orders of magnitude and suppose that sufficient time (2–3 hours) passes between two successive overloads for the transformer to return to its initial temperature. However, they provide a convenient basis to match the characteristics of a relay with those of a transformer.

Protection against short circuits

This is provided by current-operated relays associated with circuit-breakers (MV side and/or LV side) or by fuses with high rupture capacity (MV side).

It is advisable to provide equipment (tripping relays, circuit-breakers or fuses) capable of interrupting short-circuit currents in the shortest possible time.

From this point of view, fuses are preferred provided the power and voltage remain within the limits of the fuses available. In effect their characteristic under short-circuit conditions is more often such that blowing occurs within the first peak of current, which cannot therefore develop its full assymmetry. This is helpful in limiting the thermal and electrodynamic stresses on a transformer.

A standard arrangement of trip relay and circuit-breaker however can hardly interrupt a short circuit in less than 0.2 s, which allows virtually all the assymmetry to develop, but in this case the thermal stresses are limited and the electrodynamic stresses imposed on the transformer are only limited in magnitude by its own impedance and those of the system.

The choice of the short-circuit voltage of a transformer will therefore have a substantial influence on the values of the short-circuit currents, the electrodynamic forces and hence the risk of damage.

Protection against overloads

This can be carried out by thermal detectors sensitive to the maximum temperature of the dielectric liquid (called thermostats) or by a low-voltage isolator with a current operated relay. A description is also given below of winding temperature detectors (thermal images).

Thermal detectors (maximum temperature of oil)

These detectors, usually with two contacts, can actuate an alarm and trip at temperatures of 80 and 90°C which, for an ambient temperature of 20°C, correspond to stages where the contractual temperature rise of the insulation risks being, or already has been, exceeded.

Sometimes adjustments are made by taking the ambient reference temperatures as 30 or 40°C, shown in the standards as the average summer temperature or maximum ambient temperature respectively. This avoids unwanted trip outs, but introduces further risks.

In fact the time constant for the temperature rise of the transformer oil is 2–3 hours, whereas that of the windings is only 10 minutes or so, so that in the presence of a sudden large overload, the windings overheat before the thermal contacts close even when adjusted for 80–90°C at an ambient of 20°C.

Winding temperature detectors (or indicators)

By using a thermal detector similar to the preceding ones (sensitive to the maximum temperature of the oil placed in a pocket) and a heating element (resistance) through which passes a current proportional to the transformer load, an analogue display is obtained of the temperature of at least one of the windings, by simply adding the effects of the pre-existing load and the overload.

These instruments, usually used on large transformers, make it possible to operate transformers rationally and with certainty.

They are frequently called 'thermal images'. This term has been misapplied to equipment which does not take the temperature of the oil into account and can only give an imprecise image of the windings. They fall into the category of time-current integrating relays.

Relays or current sensitive trip devices

Single threshold instantaneous relays   So long as the ambient temperature does not exceed 20°C, the standard ambient temperature, a transformer can withstand prolonged overload of the order of 5–10 per cent without too much risk, and the first precaution would be to provide a relay isolating the transformer for 1.1 I_n (I_n being the rated current).

However this is too simplistic. It leads to unwanted trip outs, if the relays are on the supply side, when the transformer is switched in and it will interrupt any surge exceeding 1.1 I_n when a motor is started up, for example, whereas it is known that the transformers are quite capable of withstanding them.

Raising the threshold to 1.3 I_n will not always suffice to overcome the switching-in, and it might risk allowing an overload just less than 1.3 I_n to last too long.

Double threshold relay (with one on time delay)   The first improvement is obtained by using a double threshold relay. The first, which trips out instantaneously for 3–6 I_n, allows the switching-in peak to pass but operates on a short circuit. The second has a delayed action (0.5–5 s) adjusted for 1.1 I_n for example to provide protection against overloads of long duration, without frequently reacting to short term overloads.

Current dependent time relay   The preceding text leads one to believe that groups of relays are used to give several thresholds in which the time delay would be drawn from the characteristic $t = f(n)$ referred to earlier.

The types currently manufactured generally have an instantaneous relay for 3–6 I_n, as previous, with a time relay in which the delay depends on the current in accordance with the law of the type $t = f(1/n)$, or preferably $t = f(1/n^2)$.

In this class are found, in particular, thermal relays and (at one time) static relays, some of which no longer require an auxiliary supply and can be used with low-voltage circuit-breakers. The adaptability of static relays

Table 18.3

	n	1.2	1.5	2	3	4	5	6
Transformer		175	110	63	28	16	10	7
(1) Static relay		175	100	40	18	10	6	4
(2) Thermal relay		200	95	45	21	14	12	9.5
(3) Inverse time relay		25	15	10	6.5	5	4.4	3.8

(Characteristics $t = f(n)$)

enables a single model of a relay to reproduce a wide variety of characteristics, which allows it to be better adapted to those of a given transformer.

Table 18.3 compares the characteristics referred to earlier for a transformer and those obtainable by various types of relays currently available.

It can be seen that relay (3) which intervenes always well ahead of the characteristic time t of the transformer, can be used in several cases. However, it can be criticized for not performing as well as relay (1) in allowing optimum use of the transformer particularly in the range of slight overloads.

Relay (2) appears to be less suitable at first sight. However, this impression should be corrected by taking into account

(a) the adiabatic theorem, correct for time periods of 10 s or so, but excessively conservative above 1 minute,
(b) the action of the instantaneous trip out which can be adjusted here between 4 and 5 I_n.

This outline example demonstrates the possibilities that a detailed study of an overload protection system provides.

19 Furnace transformers

Experience has been gained over recent years in the construction of core-type transformers, especially with regard to the ability of power station auxiliary transformers of low impedance to withstand short circuits. This knowledge plus the development of high current technology (experience of high power transformers in power stations) has made it possible for reliable, economic high power core-type transformers to be manufactured for use with furnaces.

POWER REQUIREMENTS OF FURNACES

High currents

The supply voltage to furnaces, particularly arc furnaces, is generally at hundreds of volts. Since the power may be as much as 100 MVA, the currents may be of the order of tens of thousands of amperes.

Wide voltage fluctuations

The supply voltage of electric arc furnaces must be able to vary widely (in the ratio of three to one or even more), depending on the stage of operation (melting or refining). This voltage variation must be carried out continuously, at least for large power furnaces, by an on-load tap changer so as to reduce the number of times the circuit-breaker operates. The regulation is quite fine, and this requires an on-load tap changer with a large number of steps.

Figure 19.1. 80 MVA 20 000/250 to 750 V furnace transformer

Low short-circuit voltage

The impedance of the cables between the transformer and the furnace is always large, and the supply sometimes incorporates several transformation steps in series, the impedance of which must be kept low so as not to reduce the efficiency of the installation.

Demand for variable current

The demands for current vary widely, particularly during the period of melting down, with numerous short circuits at the furnace electrodes.

Over-voltages

The frequent rupture of the arc, particularly during melting down, causes over-voltages.

SUPPLY ARRANGEMENTS

Three arrangements are most commonly used, as follows:

Direct supply by variation of induction

This arrangement (see Figure 19.2) is used for low power installations up to 10 MVA where the furnace is supplied directly from 20, 30 or 60 kV. This is the simplest arrangement of all and is possible in this case because the cheap on-load tap changers of current manufacture can only be used for relatively low currents and voltages. The voltage difference from one step to the next is not, unfortunately, the same throughout the range of regulation.

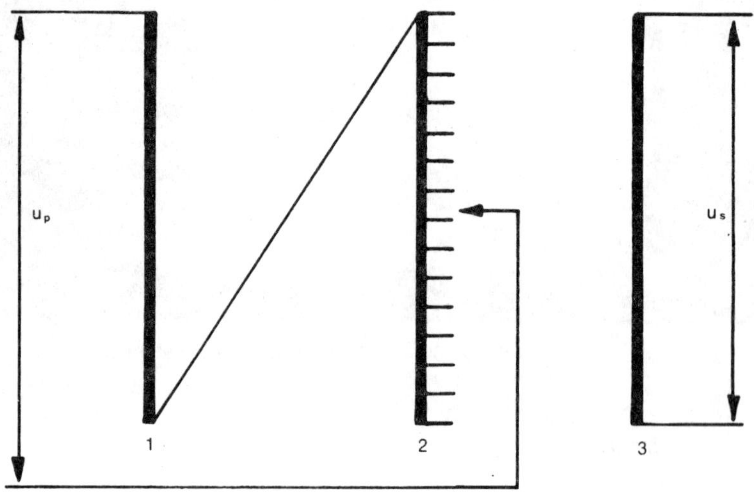

Figure 19.2. Direct supply by variation of induction. U_p Primary voltage. U_s Secondary voltage. 1. Primary main winding. 2. Primary tapping winding. 3. Secondary winding

Supply in cascade

This arrangement employs an auto-transformer for regulation and a furnace transformer (see Figure 19.3). It is used for high power furnaces

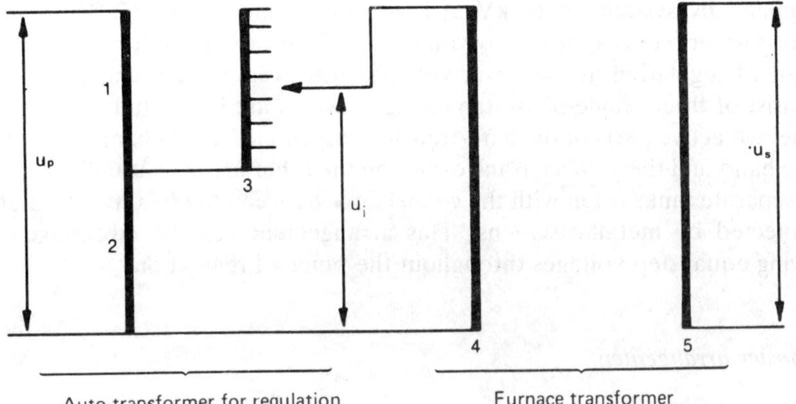

Figure 19.3. Supply in cascade. Auto-transformer for regulation and furnace transformer in a fixed ratio. U_p Primary voltage of auto-transformer. U_i Secondary voltage of auto-transformer, primary voltage of furnace transformer. U_s Secondary voltage of furnace transformer. 1. Series winding auto-transformer. 2. Shunt winding auto-transformer. 3. Tapping winding auto-transformer. 4. Primary winding furnace transformer. 5. Secondary winding furnace transformer.

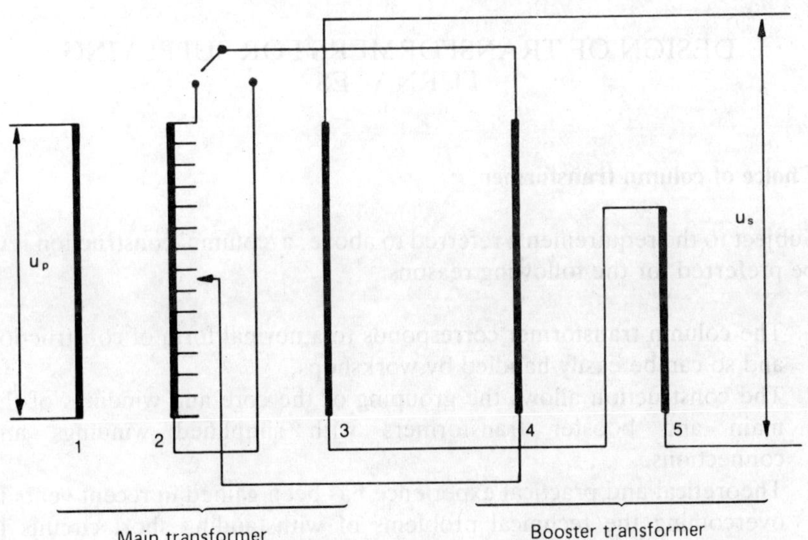

Figure 19.4. Supply by booster. U_p Primary voltage. U_s Secondary voltage. 1. Primary winding of main transformer. 2. Tapping winding of main transformer. 3. Secondary winding of main transformer. 4. Primary winding of furnace transformer. 5. Secondary winding of furnace transformer.

supplied by systems at 60 kV at least, and often at 220 kV. The auto-transformer reduces the supply voltage to the furnace transformer and uses on-load regulation at the lower voltage. However, the tap changer must consist of three single-phase tap changers which are bulky and expensive. The two active parts of the auto-transformer with their tap changers on the one hand and the furnace transformer on the other are each usually placed in separate tanks often with the connections between the two transformers protected by metallic screens. This arrangement has the advantage of giving equal step voltages throughout the range of regulation.

Booster arrangement

The arrangement shown in Figure 19.4 is normally used for medium and high power installations. Because of the intermediate winding of the main transformer, the cost of the on-load tap changer is kept as low as possible by carefully choosing the current and voltage. The variations of secondary voltage are identical from one step to another throughout the range of regulation. It is possible to place both active parts in the same tank, thereby economizing on the steel fabrication. This is a recommended arrangement used for all large furnace transformers. It is the arrangement considered in the subsequent text.

DESIGN OF TRANSFORMERS FOR SUPPLYING FURNACES

Choice of column transformer

Subject to the requirements referred to above, a 'column' construction is to be preferred for the following reasons:

1. The column transformer corresponds to a normal form of construction and so can be easily handled by workshops.
2. The construction allows the grouping of the core and windings of the main and booster transformers with simplified windings and connections.
3. Theoretical and practical experience has been gained in recent years in overcoming the technical problems of withstanding short circuits in column transformers of low short-circuit voltage (power station auxiliary transformers).
4. Technical solutions relating to high currents in large power station transformers are directly usable.

Grouping of two active parts

The two active parts, of the main and booster transformers, are placed in one tank to reduce the cost of steel fabrication, decrease the number of output terminals and the connections. Each part of this form of construction, circuit, windings, connections, terminals and accessories, is examined below.

Magnetic circuits

The circuits of each of the cores are constructed as three columns (see Figure 19.5), in low-loss grain-oriented sheet, mitred and with joints at 45°. No bolt passes through the core or the yokes. The latter are clamped between thick steel sheets by means of glass fibre bands to avoid the formation of a turn in short circuit.

Figure 19.5. Core and windings of 44 MVA furnace transformer showing the two magnetic circuits attached to the cover

To simplify the winding (see below), the two magnetic circuits have the same core height and the same centre line distance, but different diameters and cross-sectional areas. The ratings of the two active parts are quite different, the booster being sufficient for regulation only.

Windings

The HV and tap windings are standard. The LV windings, however, are quite special. Because of the heavy currents flowing through these windings and the low voltages, a small number of turns are required and parallel layer windings are used. A computer calculates the distribution of current in the different windings in parallel and hence predicts the temperature rise and the forces to which they will be subjected in short circuit. As the current is the same in the LV windings of the main transformer as in the booster transformer, the same number of layer windings with the same conductor is used in the two active parts. It is also possible to construct the windings of the two corresponding active parts in a figure of eight, thus avoiding connections between the two (Figure 19.6).

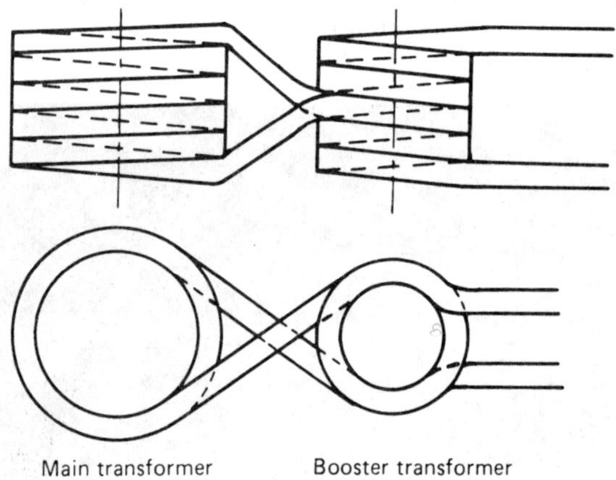

Figure 19.6. Special LV winding with common coils for the main and booster transformers

The conductor must usually have a large cross-sectional area. Transposed wire consisting of several single wires in continually changing positions is used, with the advantage of reducing supplementary losses. This construction requires the LV winding to be placed on the outside. The HV winding is placed next to the core, with the tap winding in the middle. This arrangement helps to reduce the variations of short-circuit voltage, depending on the position of the tap changer. A reversing switch on the on-load tap changer can reverse the direction of the tap winding (Figure 19.4) which enables the range of regulation to be doubled.

The windings are clamped by clamping screws with automatic take up (retraction). Each winding is fixed separately. Good performance is thus assured of all the windings during the numerous short circuits to which this type of transformer is subjected.

Connections

The HV and tap connections are standard. The LV connections however are special. It is necessary to connect in parallel a large number of coils placed one over the other, with the coil ends coming from the booster winding (see preceding section, winding in figure of eight). The windings are made in parallel in several groups, depending on the number of outlets required for the transformer–furnace connection.

As far as possible, the lengths of the connections between each group of windings and the corresponding terminals are kept the same, and the number of braised joints has to be reduced to avoid the possibility of unbalanced impedances. As the windings are placed over one another, the connections are made in a straightforward manner using vertical copper bars.

To avoid stray losses and additional temperature rise the input and output bars for each group are placed near each other. This reduces the overall magnetic field around them.

Arrangement and types of LV terminals

The parallel arrangement of the LV windings obviously affects the layout of the LV terminals on the vertical face of the tank, with a minimum length of connections, reducing the losses, temperature rises and the cost of the transformer. Nevertheless it is always possible to position the terminals on the top cover.

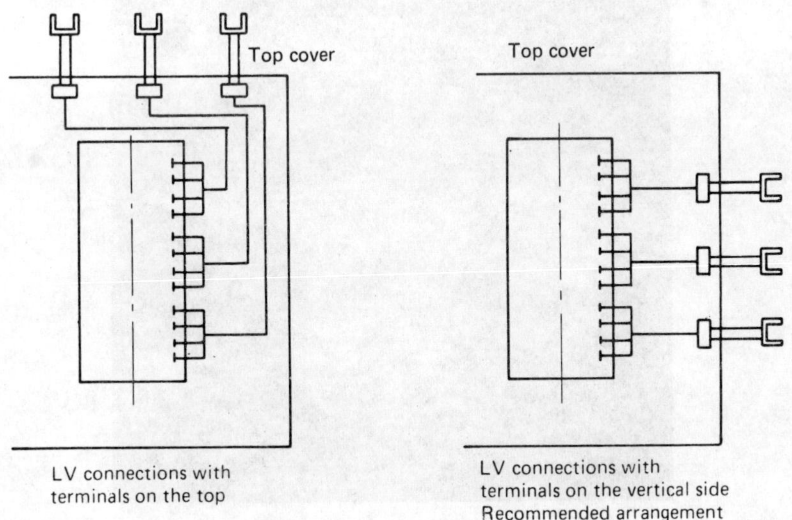

Figure 19.7. Arrangement of connections and LV terminals

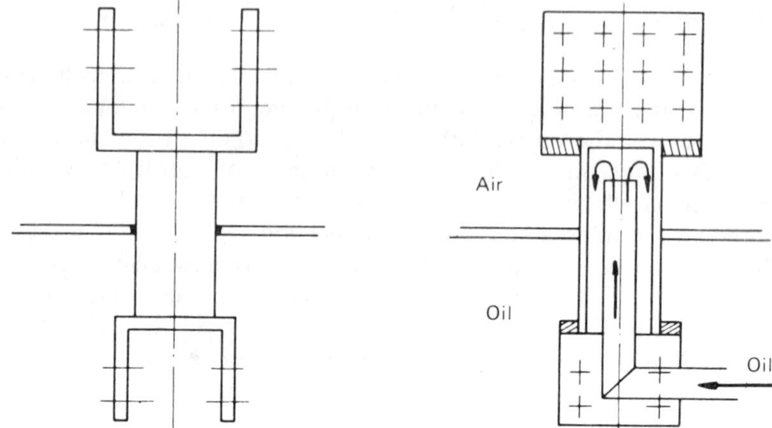

Figure 19.8. Principle of oil-cooled terminals

The terminals used on the LV side are forced oil circulation type with internal and external connections by U-shaped flags with adjustable direction (Figure 19.8).

The terminals are standardized from 12 to 40 kA. The cooling oil is gathered by suction pipework at the bottom of the tank and circulated by a pump. (Water-cooled terminals can also be used.)

The terminals are fixed in plates of insulating material which are themselves fixed on the side or the cover of the tank (Figure 19.9).

Figure 19.9. Core and windings of 44 MVA furnace transformer showing the HV connections and the LV terminals on the top

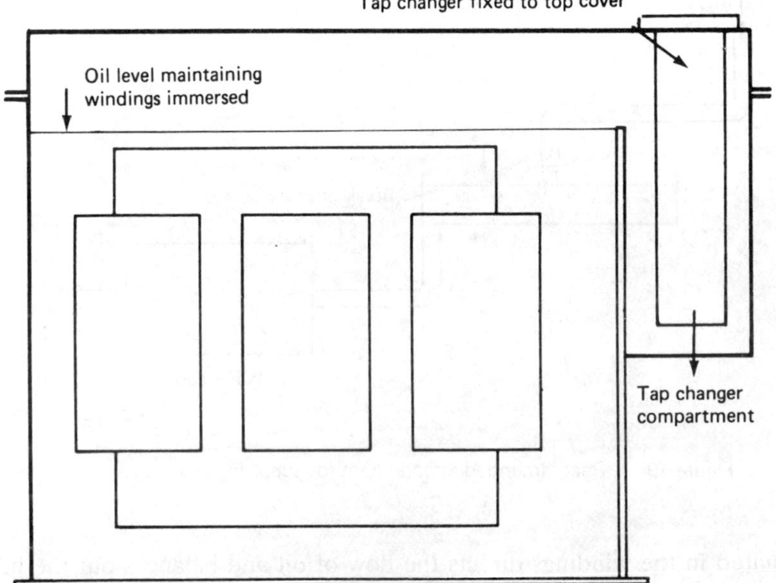

Figure 19.10. Arrangement of on-load tap changer

On-load tap changers

The booster regulation system enables a tap changer of current manufacture to be used in the majority of cases. But it must be carefully maintained because of its frequent operation. The oil in the tap changer is filtered after each operation to avoid rapid pollution, and the tap changer is placed in a special compartment making it possible to renew the contacts without draining down the oil too far and exposing the windings (Figure 19.10).

Tank and cover plate

The tank is rectangular in shape with flat cover and bottom. It has numerous stiffeners to enable a vacuum to be applied. The side or cover along which the connections pass are in non-magnetic steel to avoid temperature rise due to induced currents.

Cooling

Water coolers are usually used. The oil cooled by these coolers is pumped through all the windings of each phase of the two units. A system of baffles

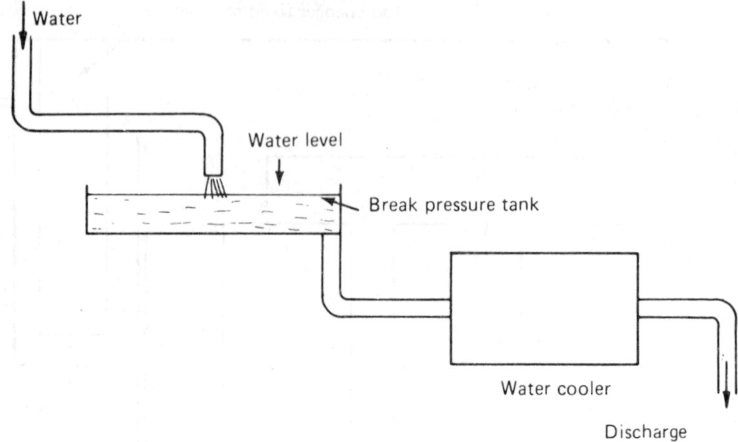

Figure 19.11. Recommended arrangement for supplying water to a cooler.

situated in the windings directs the flow of oil and balances out the head losses so that cooling of the conductors is uniform throughout.

The water coolers consist of tubes in which there is an exchange of heat between water and oil. To avoid any leakage of water into the transformer oil, the supply system shown in Figure 19.11 is advised. The pressure of water in the cooler is at a value fixed by the height of the break pressure tank above the cooler. This value can be adjusted to be slightly less than the pressure of oil in the cooler.

Figure 19.12. 44 MVA furnace transformer ready for shipment

Control and protection

The LV current is monitored by a current transformer usually placed in the tap winding of the principal transformer inside the tank.

A Buchholz relay can detect the presence of gas in the oil.

Another Buchholz relay protects the on-load tap changer. One or more thermometers check the temperature of the oil. One or more winding-temperature indicators can also be installed to check the temperature of the oil and the windings.

CONCLUSION

The column form of construction for transformers supplying furnaces, including arc furnaces, is proving to be quite suitable even for high power installations. When special arrangements are adopted for the terminals, they can also be very economic.

20 The transformer and energy saving

INTRODUCTION

In the preceding chapters, we have considered the problems of transformer construction and use, and ways in which those problems can be overcome. Of all electrical devices, the transformer is amongst the most efficient. However it does still have power losses which show in the form of heat.

Figure 20.1. 100 MVA transformer installed at a 220/90 kV substation interconnecting two systems; cooling by air coolers

In Chapter 3, it was shown that low-loss transformers can be constructed, which increase the capital cost, but are profitable in the long term, taking into account the capitalization of the losses.

This chapter is concerned with the 'Transcalor System', which is a series of methods of recovering the heat losses to raise the efficiency of the whole transformer installation to almost 100 per cent in some conditions. While this chapter may appear to be the antithesis of Chapter 3, Capitalization of Losses, it is really complementary. To use a transformer as a resistance is absurd, particularly as its aim is to transform electric current with the least possible losses.

Accurate up-to-date information is difficult to obtain, so use will be made of more reliable information from past years. Some calculations will, however, be approximate but quite adequate for giving an order of magnitude.

ENERGY LOSSES AT NATIONAL LEVEL

The consumption of energy in France in 1979 in megatonnes equivalent petroleum was as follows:

By energy source		By sector	
Coal	32	Industry	60.5
Petroleum	111	Domestic	68
Natural gas	23	Transport	36.5
Hydro power	14	Energy	24
Nuclear	9		
	189		189

The consumption of electricity was divided between the high voltage and low voltage sectors in billions of kWh as follows:

High voltage	157
Low voltage	81
	238

The production of this power required about 53 million tonnes of equivalent petroleum (megatep) using the usual ratio of

1 billion kWh = 220000 tep

12 million tonnes of which are crude petroleum.

Transformer losses

The number of transformers installed in France by 1979 can be grouped into the sizes shown in Table 20.1.

Table 20.1

Type of Transformer	Approximate total installed power (MVA)	Average losses at 70 per cent load (per cent)	Average losses (MW)
Power station transformers	40 000	0.2	80
HT System transformers	60 000	0.2	120
Industrial transformers	40 000	0.4	160
63 and 90 kV/MT transformers	40 000	0.3	120
MT/LT transformers	120 000	0.4	480
TOTAL	300 000		960

The overall losses at national level are therefore equivalent to the output of one nuclear power station. Moreover, during a year, allowing for the operating conditions, they account for the consumption of 3.2×10^6 MWh or 32 billion kWh which at the previously mentioned conversion ratio is $3.2 \times 220\,000 = 704\,000$ tep. This represents 0.7 per cent of the annual consumption of petroleum, and 6 per cent of the tonnage of petroleum necessary to produce the energy usefully consumed. These percentages are low but significant when the absolute value of the 704 000 tep is taken into account.

THE ECO-ELEC ENQUIRY

In June 1980, a report was published by the agency for energy economy under the title 'Eco-Elec' describing the work done by a working party in which EDF, APAVE, FIEE and FIMTM participated. Its aim was to define the methods to be used in industry to achieve the stated objective of saving electricity.

The analysis

Many types of electrical equipment or sectors were examined, in particular transformers, capacitors, cables, motors, electrical heating and lighting. It was concluded that if it would be possible to achieve a valid level of saving on capacitors and motors in particular, the saving of energy to be expected from transformers would be of the order of 80 000 tep/annum after substantial capital investment. This was not regarded as a useful investment in relation to other types of equipment examined, motors for example, for which the saving would be of the order of 1 million tep/annum.

The findings

In studying the potential for energy saving, the working party took as the basis the replacement of old model transformers by modern ones with lower losses.

A saving of 3 per cent could be expected on the total energy consumed for designs predating 1935 and 2 per cent for models predating 1950.

However, taking into account the number of transformers involved, about one third of the total but representing only one sixth of the installed power, it was concluded that the gain was insignificant, of the order of the 80 000 tep already mentioned.

It is clear that the transformer is already a very efficient piece of equipment. Any improvement is therefore difficult to achieve.

Moreover many transformers constructed in the 1950s are still in good condition, which hardly encourages their replacement for just a small improvement in performance and an uncertain return on the investment. This does not apply when increased power is required however, as it is often more economical to completely replace an old transformer by a more powerful one than to add a second to an existing.

Without being able to act effectively on the causes, one can try to act on the effects, i.e. recover the heat dissipated as pure losses. This is the origin of the Transcalor System.

THE RECOVERY OF LOSSES

This appears to be a philosophy which, at first sight, runs counter to that proposed in Chapter 3, namely, the promotion of the construction of low loss transformers, because if most of the losses can be recovered, it is not clear how the additional cost of buying a low loss transformer can be justified.

This reasoning would be false, even if the losses recovered could be used effectively, because taken to the limit it would correspond to the use of the transformer as a heating element, which is obviously a nonsense. In fact it usually makes sense to buy a low loss transformer even if parallel recovery raises the efficiency of the installation to almost 100 per cent. Where a transformer of normal loss level exists, the recovery of heat is obviously even more important.

TECHNICAL ASPECTS OF RECOVERY

There are two possible basic solutions, namely to extract the heat from the transformer cooling fluid directly with the assistance of an exchanger, or to extract the heat indirectly by means of an exchanger and a heat pump. The second method has been called the Transcalor System following a study made by Alsthom in 1979. The choice of method depends on numerous conditions as will be explained later.

Direct recovery by exchanger

The installation is shown in Figure 20.2. Such an arrangement is not new, and has already been used. It is employed by the Pechiney Company, at Venthon, where a saving of 130 tons of fuel per year was immediately made by recovering about 500 kW of heat from three transformers. This solution is simple as it uses only static elements, but it has its limitations.

Thermally all it does is to transfer heat from one fluid to another, within the efficiency of the exchanger, and at the same temperature.

Therefore, if it is intended to use the transformer as the only source of heat without an additional heater, the temperature of the water in the circuit must be significantly high. This supposes that the transformer is sufficiently and continuously loaded, because at half load the temperature rise of the oil is from 20° to 30°C only, which is difficult to use for heating purposes taking into account the surface area of existing radiators.

Moreover, if the place to be heated is some distance from the transformer, the heat losses must be taken into account, as the water is transferred at its high temperature at which the losses are highest.

Nevertheless, it is an economic and useful system in the right conditions, as was the case at the Venthon factory for the continuous refining of aluminium. It should be noted that the pressure upstream must always be higher than that downstream to avoid entry of water into the transformer. For the same reason plate exchangers should preferably be used.

The Transcalor System

We have seen that the direct recovery by exchanger requires at least two conditions to be met: a high transformer load and a continuous load. The Transcalor System is not subject to these two conditions. In this system, a heat pump is inserted in the circuit downstream of the exchanger and upstream of the point of use (see Figure 20.3). By coincidence both Electricité de France (EDF) and the private sector had the same idea, without any prior discussion, for an installation at the EDF–CRTT Paris-Group, operated by Nord Ouest-Puteaux, on the initiative of the operating organization.

A 100 MVA Alsthom transformer is used as the source of heat. To our knowledge it is the first time such a system has been used in France.

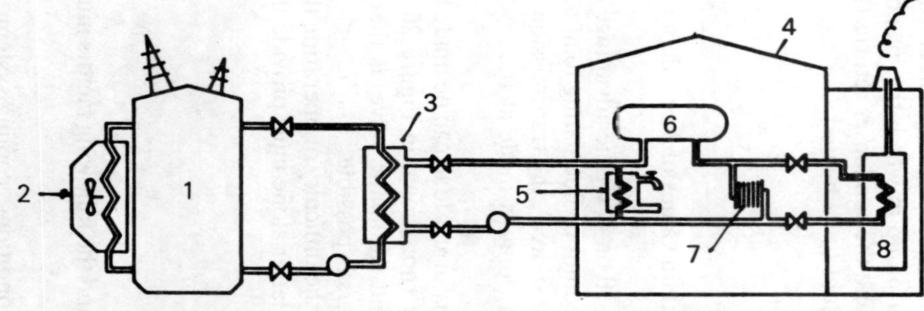

Figure 20.2. Simplified installation. 1. Transformer. 2. Air coolers. 3. Heat exchanger. 4. Building to be heated. 5. Domestic hot water. 6. Hot water storage tank. 7. Radiators for space heating. 8. Hot water heater (not used at same time)

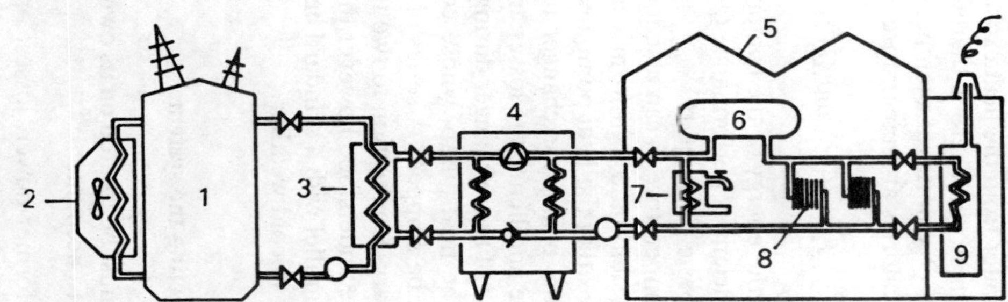

Figure 20.3. Transcalor System with heat pump. 1. Transformer. 2. Air coolers. 3. Heat exchanger. 4. Heat pump (water/water). 5. Building to be heated. 6. Hot water storage tank. 7. Domestic hot water. 8. Radiators for space heating. 9. Hot water heater (not used at same time)

The advantage of the solution

The principle of the heat pump is well known. It is a machine whose thermodynamic circuit uses a transfer fluid to extract heat from a source, called the 'cold source', and return it in a more usable form to the place to be heated called the 'hot source', using, in the process, a certain amount of energy in the compressor.

In the case under consideration, the cold source would be the circulation of cooling water from the heat exchanger, which in turn receives heat from the transformer oil. There is therefore a heat pump in which the evaporator extracts Q_1 calories at a temperature T_1, and the condenser returns the heat as Q_2 calories at temperature T_2 to the fluid to be heated.

$$Q_2 > Q_1 \quad T_2 > T_1 \quad \text{and } Q_2 = Q_1 + W$$

where W is the energy of the compressor.

The coefficient of performance Q_2/W can attain values of 3 or 4 depending on the basic parameters.

As the temperature T_2 can be equal to $T_1 + 30$ or 40°C, the advantage of using the system on a transformer at low, medium or fluctuating load can be seen. By siting the heat pump near the 'hot source', the transmission of heat from the primary exchanger to the heat pump is made at the lower temperature and hence the losses are lower.

Direct use of the exchanger during high transformer load can be made by bypassing the heat pump, whose compressor (not shown in Figure 20.3) would then be stopped. A series of probes and automatic valves is therefore necessary. Connection to two transformers is possible.

It must be understood however that due to the difficulty of operating the system in parallel with a standard heater, the latter alone must provide the heat supply in cold weather.

Factors affecting the system

Each case must be treated on its own merits, and the following factors must be taken into account:

- the transformer power, losses, operating conditions, cooling system,
- the requirement for heat, for what use, in what form,
- layout of the site,
- radiator capacity when heating a building,
- climatic conditions,
- thermal insulation of the buildings.

The following typical cases will show what can be expected in terms of the energy balance and in economic terms.

THE HEAT BALANCE OF THE TRANSCALOR SYSTEM

In Table 20.2 the following are assumed:

1 kW = 860 kcal/h,
1 kW h = 0.1 kg of petroleum with calorific value of 8600 kcal/kg,
transformer load: 70 per cent for 2500 hours per year,
exchanger efficiency: 90 per cent,
efficiency of fuel oil heater: 70 per cent,
coefficient of performance of heat pump: 3,
heating ratio: 1 kW for 30 m^3.

Table 20.2 Balance with the transcalor system

Power (MVA)	Recovery per hour (kcal)	Volume heated (m^3)	Saving (tep/annum)
5	20 000	850	9
10	38 000	1 440	15
20	60 000	2 400	25
40	112 000	4 500	42
100	225 000	8 800	83

Analysis of the table

It can be seen immediately that transformers below 5 MVA are not worth
considering. If it is accepted that the height, floor to ceiling, of a building is

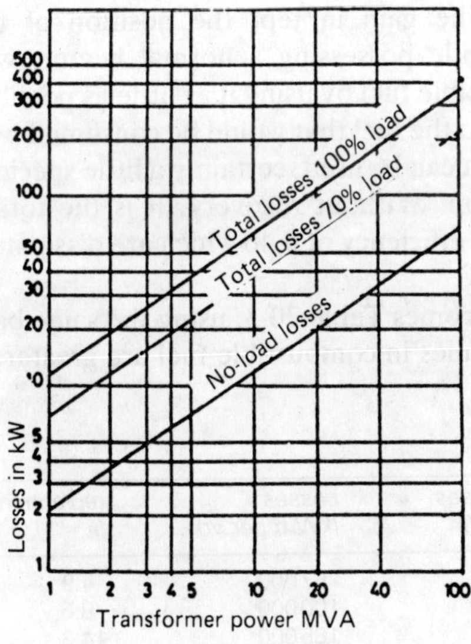

Figure 20.4. Transformer losses against power

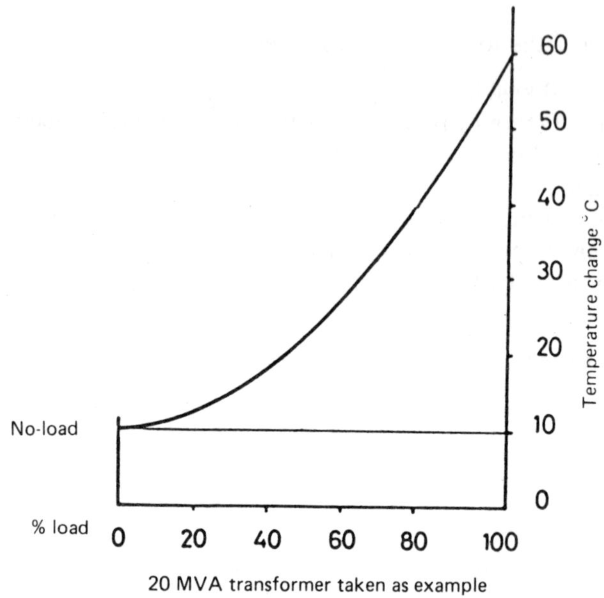

Figure 20.5. Oil temperature rise against load

2.50 m, a volume of 850 m³ represents an area of only 340 m² which is small for an industrial building, although significant for an office.

With regard to the gain in tep, the position of the industrialist is considered who, while possessing a normal heater, wishes to make an economy in combustible fuel by using it as little as possible. This is why the gain is converted into the fuel that would be consumed by using the heater.

Another argument can be used, certainly a little specious, by considering the national situation. Without recovery, it is the total kW supplied by EDF with an overall efficiency of 0.46 which are dissipated as pure losses in the environment.

The table then becomes Table 20.3, using the same basic data. It can be seen that the economies in combustible fuel are greater.

Table 20.3

Transformer power	Losses (kW)	Losses (kW h per yr)	tep/annum ($\eta = 1$)	tep/annum ($\eta = 0.46$)
5	23	57 000	4.9	11.3
10	40	100 000	8.6	20
20	66	165 000	14.3	33
40	110	275 000	24	56
100	220	550 000	48	112

In Table 20.2, heavy fuel was assumed as it is the cheapest. Clearly if the consumer burns light fuel, which is much more expensive, the saving will be increased by 70–80 per cent.

The thermal balance of the direct recovery system is not given, but the values in the two last columns of the tables should be reduced by 25–30 per cent.

ECONOMIC ASSESSMENT OF THE TWO SYSTEMS

Two methods of assessment are used. The first makes an assessment at the end of ten years as a function of the power. The second gives the time for return of investment, also as a function of the power. The assumptions are based on average values.

In both cases the rate of interest on the capital invested in the installation over the time considered is 12 per cent and annual inflation in the value of money, or increase in the price of petroleum is 10 per cent.

It is assumed that heating radiators exist already, and the installation is made by the user himself.

First method: over a period of ten years

Ten years has been taken because it corresponds to the service life of the heat pump without any overhaul.

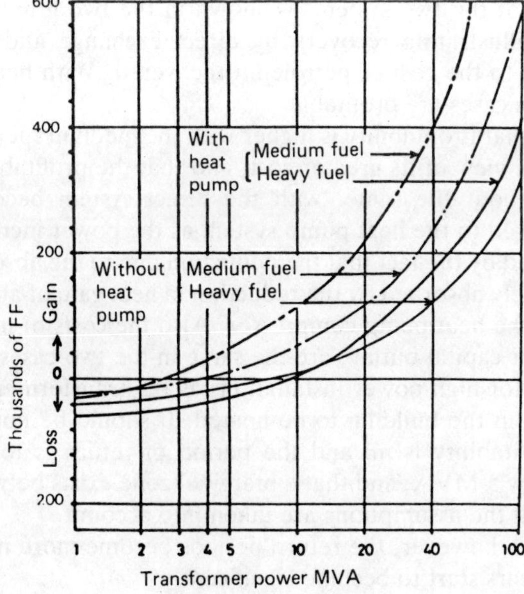

Figure 20.6. Capitalization of losses (see Chapter 3)

Second method: time for return of investment

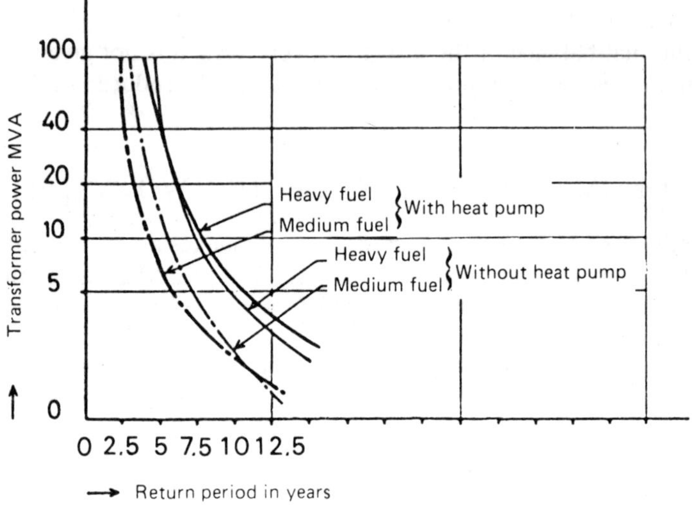

Figure 20.7. Maximum use of losses

Analysis of the two graphs

The profitability of the two systems is shown in the two graphs (Figures 20.6 and 20.7), illustrating recovery by direct exchange and with heat pump, compared to the cost of petroleum recovered. With heavy fuel or medium fuel both cases are profitable.

It can be seen that profitability is higher with the medium fuel oil, which is quite normal in view of its greater cost, and that the profitability of the two systems is about the same, with the direct system becoming less profitable compared to the heat pump system as the power increases.

This is explained by the fact that the deduction due to the absence of the heat pump is largely absorbed by the reduction in heat gain of about 30 per cent brought by the heat pump compressor. Also the costs of installation, which are a major capital outlay, are the same in the two cases. They are particularly high for high power installations as the transformer is usually some distance from the building to be heated. It should be noted, above all, that the profitability is nil and the period of return is too long for installations below 5 MVA, and that a marginal zone exists between 5 and 10 MVA when all the assumptions are taken into account.

Above 10 MVA however, the return periods become more normal and the gains at 10 years start to become significant.

CONCLUSION

There are three principal requirements for the recovery of heat losses from transformers:

- a demand for heat,
- a source of heat sufficient to meet the demand,
- a positive economic return compatible with the return currently used for the appraisal of capital investment projects within the organization concerned.

The choice between the two systems, direct or with heat pump, is made at technical level essentially as a function of the transformer operating conditions.

The economic balance at the end of the period is generally in favour of the system with the heat pump, as the gain in heat largely compensates for the cost of the heat pump. The two choices are clear:

1. If the load is high and constant, and a low capital cost is required, a simple exchanger will be chosen.
2. If the load is low or fluctuating and a long term profitability is required after making the necessary capital investment, then the 'Transcalor System' will be chosen.

If the two assessments are combined it will be seen that they both show a zone of nil return below 5 MVA, a marginal zone between 5 and 10 MVA depending on the installation conditions, and a zone of increasing return above 10 MVA.

In conclusion and without prejudging the future it may be assumed

- that the price of petroleum will increase steadily,
- that the cost of heat pumps will fall as foreign competition and mass production take effect (there are already signs of this happening),
- that interest rates will return to more normal values.

These factors will make the recovery of heat from transformers even more attractive, perhaps for uses other than heating buildings which is a limited example because of its seasonal nature.

Bibliography

Chapter 1

R. DIDES: Les efforts électrodynamiques dans les transformateurs à colonnes. *Techniques CEM*, No. 100–101, Dec. 1977, pp. 34–39.

R. ROUSSELIN et J. BASSOT: Le séchage en phase vapeur des transformateurs. *Techniques CEM*, No. 92, Dec. 1974, pp. 2–8.

Y. MARTIN: Laboratoire H.T. du Havre: progrès techniques des transformateurs. *Techniques CEM*, No. 81, July 1971, pp. 41–45.

Chapter 2

Norme Française NF C 52–100: Power transformers – Requirements (June 1970).

GAUSSENS-CAZALET: Le réglage de la tension dans les réseaux de distribution. *Bulletin de la Société Française des Électriciens*, July 1960, pp. 478–496.

PARDIGON-MESTRES: Réseaux de distribution. Techniques de l'Ingénieur – Section Electrotechnique.

MASCHINENFABRIK REINHAUSEN: Regensburg. R.F.A. On load tap changers. Service Instructions, Voltage Regulator MK10.

LE BOURGEOIS: Le poste d'alimentation de la Société des Aciéries de Lorraine. *Techniques CEM*, No. 74, Sept. 1969, pp. 18–26.

Chapter 3

E.T. NORRIS: Capitalization of technical performance. *Proceedings of the Institution of Electrical Engineers*, 1946, Part II, pp. 137–139.

P.-G. LAURENT: Considérations sur le dimensionnement optimum économique des transformateurs de distribution publique. *Bulletin de la Société Française des Électriciens*, Sept. 1955, pp. 637–651.

W. HEINIGER: L'influence de la capitalisation des pertes sur la construction des grands transformateurs. *Revue Brown-Boveri*, No. 11–12, Dec. 1965, pp. 804–808.

CONGRÈS UNIPÈDE-LA HAYE 1973: Rapport 71-03. Caractéristiques des fournitures d'énergie électrique à quelques grandes industries.

Elec. de France.: Journées d'informations électro-industrielles – Elec. de France – Paris, Nov. 1974.

Chapter 4

Norme Française NF C 52–100: Power transformers (Suppl. D & E).

Le fonctionnement en parallèle des transformateurs – KUBLER. *Revue Brown-Boveri*, Oct. 1916, pp. 154–159; Nov. 1916, pp. 171–177; Dec. 1916, pp. 195–196.

Les essais des transformateurs industriels – LAPINÉ (Dunod).

Elektrische Maschinen – III – Die transformaten – RICHTER (Springer), 1931.

Elec. de France: Journées d'informations électro-industrielles – Elec. de France – Paris, Nov. 1974.

Chapter 5

GOTTER: Erwärmung und Kühlung elektrischer Maschinen. (Springer), 1954.

McADAMS: Heat transmission – McGraw Hill

IEC: Publication 76. Power transformers

IEC: Publication 85. Thermal evaluation and classification of electrical insulation.

Norme Française NF C 52–100: Power transformers.

Norme Française NF C 51–100: Machines électriques tournantes.

DOBSA: Le refroidissement des grands transformateurs. *Revue Brown-Boveri*, 1967, No. 4, pp. 33–39.

Chapter 6

MONTSINGER: Loading Transformers by Temperature. *Journal of the AIEE*, 1930, pp. 293–297.

FABRE: Les lois de dégradation du papier imprégné d'huile dans les transformateurs. *Bulletin de la Société Française des Électriciens*, 1959, No. 103, pp. 409–418.

FAVEZ: Considérations générales sur la détermination des règles de charge des transformateurs. *Revue Genérale de l'Électricité*, July–Aug. 1969, pp. 749–753.

SALGUES: Définition et justification des limites thermiques prises en compte pour fixer les règles de charge des transformateurs d'interconnexion et de livraison à la distribution d'E.D.F. *Revue Générale de l'Électricité*, July–Aug. 1969, pp. 754–762.

LARRUE: L'application pratique des règles de charge des transformateurs d'interconnexion ou de livraison aux réseaux de distribution, *Revue Générale de l'Électricité*, July–Aug. 1969, pp. 763–768.

IEC: Publication 354. Loading guide for oil immersed transformers

Chapter 7

COPPADORO: Autotransformatori di grande potenza per l'interconnessione di réti ad altissima tensione. *L'Elettrotecnica*, 1967, No. 1, pp. 21–35.

SALGUES: Problèmes de tenue au court-circuit des auto-transformateurs. *Bulletin de la Société Française des Electriciens*, June 1964, pp. 335–343.

MONNET: La répartition de la tension dans les auto-transformateurs lors des défauts sur les réseaux raccordés. *Bulletin de la Société Française des Électriciens*, June 1963, pp. 344–351.

COPPADORO: Le réglage de tension des transformateurs de redresseur au silicium. *Revue Brown-Boveri*, 1972, No. 8, pp. 416–421.

Chapter 8

DEMIERRE: La protection des transformateurs. *Revue Brown-Boveri*, Nov.–Dec. 1966, pp. 820–827.

GUIOT: Protection contre les surtensions. *Techniques CEM*, No. 83, Feb. 1972, pp. 31–36.

CHRISTOFFEL: L'influence de tronçons de câbles sur les phénomènes de surtension des systèmes de transmission à moyenne et haute tension. *Revue Brown-Boveri*, June 1964, No. 6, pp. 369–376.

Relais de protection Buchholz type C. Notice FC 150, April 1973, Alsthom (Savoisienne).

Norme Française NF C 13–200: Installations électriques à haute tension. Dec. 1974, Part 5: Choix et mise en œuvre des matériels.

Chapter 9

Norme Française NF C 10–100: Matériel pour réseaux à courant alternatif à H.T. Coordination des isolements – Règles (Sept. 1977).

Norme Française NF C 12–100: Textes officiels relatifs à la protection des travailleurs dans les établissements qui mettent en œuvre des courants électriques (Feb. 1975).

Norme Française NF C 13–100: Postes de livraison établis à l'intérieur d'un bâtiment et alimentés pár à un réseau de distribution public de deuxième catégorie – Règles de construction et d'installation (June 1983).

Norme Française NF C 13–200: High voltage installations – Rules (Dec. 1974).

Norme Française NF C 52–100: Power transformers – Requirements (June 1970).

Elec. de France – CENTRE D'ÉQUIPEMENT DU RÉSEAU DE TRANSPORT: Cahiers de spécifications et conditions techniques relatifs à la construction des postes et de leurs annexes.

CAKEBREAD, DELIS, KIWIT, TOMATIS: Adaptation des postes à leur environnement à la fois dans les zones urbaines et rurales, y compris les problèmes du bruit et de la pollution par l'huile du sous-sol. CIGRE 1972, rapport 23-07.

Chapter 10

Calcul de la ventilation des stations de transformation. *Revue Brown-Boveri*, Jan.–Feb. 1919, pp. 9–15.

McADAMS: Heat transmission – McGraw Hill.

WILKINS, MUSA: Cooling transformers installed indoors. *Electrical Times*, 18 April 1974, p. 5.

Formulaire Pont-à-Mousson, 1975. *Hydraulique – Aéraulique*, p. 210 et seq.

Techniques de l'Ingénieur: Généralité A 700 mécanique des fluides. Mécanique et Chaleur B 230 conditionnement de l'air.

Chapter 11

M. KRONDL, E. KRONAUER: Contribution au problème du bruit des transformateurs. *Bull. Oerlikon*, 1963, No. 356, pp. 1–15.

Norme E.D.F. HN 52–02 (April 1965): Code d'essai pour la mesure dans l'air des bruits des transformateurs électriques et de leurs organes de refroidissement.

G. SPILLMANN: Lautstärken von Transformatoren und Massnahmen zur Verminderung der Geräuschabstrahlung. *BBC-Nachrichten*, 1968, No. 1, pp. 34–40.

A. COQUARD (E.D.F.): Le bruit des transformateurs: ses paramètres, les moyens de le réduire à sa source. *Revue Générale de l'Electricité*, Feb. 1969, pp. 179–194.

Chapter 12

Norme Française NF C 52–100: Power transformers – Requirements (June 1970).

Y. MARTIN Laboratoire Haute Tension du Havre: Progrès technique des transformateurs. *Techniques CEM*, No. 81, July 1971, pp. 41–46.

KRAAIJ: Progrès réalisés dans la technique de l'essai des transformateurs. *Revue Brown-Boveri*, April.1967.

CHRISTOFFEL: Les effets des nouvelles recommandations pour la coordination de l'isolement sur les conditions futures d'essai des transformateurs. *Revue Brown-Boveri*, Aug. 1972, pp. 395–398.

HUBER: Méthodes de mesure utilisées dans l'essai d'échauffement des transformateurs et ordre de grandeur des erreurs correspondantes. *Revue Brown-Boveri*, April 1967.

ROGE, PIRKTL: Problèmes relatifs à la résistance au court-circuit des gros transformateurs. *Revue Brown-Boveri*, Aug. 1972, pp. 404–409.

Chapter 13

IEC: Publication 270 (1981). Partial Discharge Measurements.

B. FALLOU, J.F. MOREL: Detecting and Locating Partial Discharges with Ultrasonic Techniques. *Revue Générale de l'Electricité*, March 1971.

P. MORO, J. POITTEVIN: Locating Partial Discharges in Transformers by Ultrasonic Detection. *Revue Générale de l'Electricité*, January 1978.

R.T. HARROLD: The Relationship between Ultrasonic and Electrical Measurement of Under-oil Corona Sources. *IEEE Transaction on Electrical Insulation*, March 1976.

Chapter 14

Norme Française NF C 27–101 (Sept. 1982): Huiles minérales isolantes neuves pour transformateurs et appareillages du connexion. Règles.

Norme Française NF C 27–221 (July 1974): Méthode pour la détermination de la rigidité diélectrique des huiles isolantes.

Norme Française NF C 27–222 (July 1974): Maintenance et surveillance des huiles isolantes en service.

Norme Française NF C 27–475 (June 1975): Méthode d'échantillonnage des diélectriques liquides.

DELÉHAYE: L'entretien des transformateurs en cours d'exploitation. *Bulletin d'information des Centrales électriques*, No. 65, July 1969, pp. 21–28.

SCHOBER: Recommandations pour le contrôle des huiles isolantes en service. *Revue Brown-Boveri*, 1972, No. 8, pp. 422–426.

VORWERK: The maintenance of substations in technical, economical and organizational respect. *Electra*, Jan. 1974, pp. 89–106.

DORNENBURG, STRITTMATTER: Surveillance des transformateurs dans l'huile par analyse de gaz. *Revue Brown-Boveri*, May 1974, pp. 238–247.

Mrs FALLOU: Detection and research for the characteristics of an incipient fault from analysis of dissolved gases in the oil of an insulation. *Electra*, Oct. 1975, pp. 31–52.

Chapter 15

W. ERB, J. SCHOBER: Recommandations pour le contrôle des huiles de transformateurs en service. *Revue Brown-Boveri*, No. 8, 1972, pp. 423–426.

E. DÖRNENBURG, W. STRITTMATTER: Surveillance des transformateurs dans l'huile par analyse de gaz. *Revue Brown-Boveri*, No. 5, 1974, pp. 238–247.

S. AUSTEN STIGANT, A.C. FRANKLIN: *J. and P. Transformer Book* (Butterworths London).

Chapter 16

Norme Française NF C 52–100: Power transformers (June 1970).
R. DIDES: Les efforts électrodynamiques dans les transformateurs à colonnes. *Techniques CEM*, No. 100–101, Dec. 1977, pp. 34–39.

Chapter 17

Coordination des isolements. Publications 71.1 et 71.2. IEC (International Electrotechnical Commission).
Coordination des isolements. Norme UTE C 10–100 (Union Technique de l'Electricité).
Matériel de protection – Parafoudres à résistance variable: Règles. Norme UTE C 65–100.
BEAUMONT, BOILEAU, RIVET: Éclateurs pour moyenne tension à dispersion réduite. *Revue Générale de l'Électricité*, No. 9, Sept. 1972.
Notices de parafoudres *Brown-Boveri*.
T.H. SIE: Rigidité longitudinale des transformateurs. *Revue Brown-Boveri*, No. 7, July 1976, pp. 465–470.
Dr CHRISTOFFEL: Les effets des nouvelles recommandations pour la coordination de l'isolement sur les conditions futures d'essais des transformateurs. *Revue Brown-Boveri*, No. 8, Aug. 1972, pp. 395–398.
E. SARBACH: Influence exercée par la valeur instantanée de la tension de service sur la tension d'amorçage au choc d'un parafoudre. *Revue Brown-Boveri*, No. 6, June 1964, pp. 358–362.

Chapter 18

B. ANGUIS: Le Kit 25 et les réseaux d'aujourd'hui, *Techniques CEM*, No. 99, pp. 17–25.
Y. POURCIN, Y. SACHER: Sécurité d'alimentation en énergie Électrique. *Revue Générale de l'Électricité*, Vol. 87, No. 1, Jan. 1978, pp. 36–42.
Norme Française NF C 64–400: Appareillage sous enveloppe métallique pour courant alternatif de tensions assignées supérieures à 1 kV et inférieures ou égales à 72,5 kV (Oct. 1984).
Norme E.D.F. HN 64 S 41: Appareillage à haute tension sous enveloppe métallique. Tableaux MT pour postes MT/BT.
C. REMOND: Le régime du neutre et ses applications dans les grands ensembles. *Revue Générale d'Électricité*, Vol. 87, No. 1, Jan. 1978, pp. 43–56.
Norme Française NF C 15–100: Low voltage electrical installations (Feb. 1981).
J. SCHMELTZ: Les prises embrochables à moyenne tension. *Le Moniteur de l'Electricité*, No. 311, Feb. 1975.
A. SCHLEICH: Le comportement des enroulements partiellement entrelacés de transformateurs soumis aux tensions de choc. *Bulletin Oerlikon*, No. 389–390, June 1969, pp. 12–20.
T.H. SIE: Essais de tenue des transformateurs aux surtensions de manœuvre. *Bulletin Oerlikon*, No. 383–384, Oct. 1968, pp, 20–26.
E. KOHLER, J. BAUMANN: Sur la disposition des parafoudres dans les postes électriques. *Revue Brown-Boveri*, No. 6, June 1964, pp. 363–368.
Techniques of high voltage testing. Part I: Definitions and procedures for tests. Publication 60-1-1973 IEC.
PAUTHENET: Les phénomènes de ferro-résonance. Cours professé à l'Institut Polytechnique de Grenoble.
E.I. DOLAN, D.A. GILLIES, E.W. KIMBARK: Ferro-resonance in a transformer switched with an EHV Line. IEEE PAPER 71 TP 533 PWR Summer meeting, July 18–23/1971.
R.W. ALFORD, J.H. HARLOW: Ferro-resonance a compendium for underground distribution systems. *Proceedings of the American Power Conference*, Vol. 34, 1972, pp. 1089–1096.

Chapter 19

G. SCHEMEL: Transformateur de fours à arcs pour les nouvelles séries de fours AM et AL. *Revue Brown-Boveri*, No. 7, 1979, pp. 446–450.

P. BONIS: Technologie des fours à arcs. CESSID, 1975.

Jean GUIGLIO (Ugine Aciers), Pierre CHASSAIN, Camille STERBA (Société Laborde & Kupfer). Alimentation électrique des fours de métallurgie. *Journal du four électrique*, No. 9, Nov. 1976.

R. DIDES: Les efforts électrodynamiques dans les transformateurs à colonnes. *Techniques CEM*, No. 100–101, Dec. 1977, pp. 34–39.

Chapter 20

S. CAGNIOUX: Rapport d'étude interne CEM 'Transcalor System', No. 79054 Dec. 1979.

Revue 'Jalons', Pechiney-Ugine-Kuhlman, No. 22, April 1979.

Dossier EDF – CRTT, Puteaux, Oct. 1981.

Index